MATH WORD PROBLEMS

DEMYSTIFIED

Demystified Series

MATH WORD PROBLEMS

DEMYSTIFIED

ALLAN G. BLUMAN

McGRAW-HILL

New York   Chicago   San Francisco   Lisbon   London
Madrid   Mexico City   Milan   New Delhi   San Juan
Seoul   Singapore   Sydney   Toronto

The **McGraw·Hill** Companies

Copyright © 2005 by The McGraw-Hill Companies, Inc. All rights reserved. Printed in the United States of America. Except as permitted under the United States Copyright Act of 1976, no part of this publication may be reproduced or distributed in any form or by any means, or stored in a data base or retrieval system, without the prior written permission of the publisher.

3 4 5 6 7 8 9 0   DOC/DOC   0 1 0 9 8 7 6 5 4

ISBN 0-07-144316-9   *32082025 6/05*

The sponsoring editor for this book was Judy Bass and the production supervisor was Pamela A. Pelton. It was set in Times Roman by Keyword Publishing Services Ltd. The art director for the cover was Margaret Webster-Shapiro; the cover designer was Handel Low.

Printed and bound by RR Donnelley.

McGraw-Hill books are available at special quantity discounts to use as premiums and sales promotions, or for use in corporate training programs. For more information, please write to the Director of Special Sales, McGraw-Hill Professional, Two Penn Plaza, New York, NY 10121-2298. Or contact your local bookstore.

 This book is printed on recycled, acid-free paper containing a minimum of 50% recycled, de-inked fiber.

Library of Congress Cataloging-in-Publication Data

Bluman, Allan G.
    Math word problems demystified / Allan G. Bluman.
        p.   cm.
    Include index.
    ISBN 0-07-144316-9 (acid-free paper)
    1. Word problems (Mathematics)—Outlines, syllabi, etc.   2. Problem solving—Outlines, syllabi, etc.   I. Title.

QA63.BS8   2005
510′.76—dc22   2004055198

To Betty Claire, Allan, Mark, and all my students who have made my teaching career an enjoyable experience.

CONTENTS

PREFACE

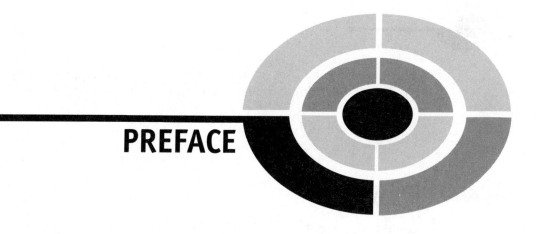

What did one mathematics book say to another one?
"Boy, do we have problems!"

All mathematics books have problems, and most of them have word problems. Many students have difficulties when attempting to solve word problems. One reason is that they do not have a specific plan of action. A mathematician, George Polya (1887–1985), wrote a book entitled *How To Solve It*, explaining a four-step process that can be used to solve word problems. This process is explained in Lesson 1 of this book and is used throughout the book. This process provides a plan of action that can be used to solve word problems found in all mathematics courses.

This book is divided into several parts. Lessons 2 through 7 explain how to use the four-step process to solve word problems in arithmetic or prealgebra. Lessons 8 through 19 explain how to use the process to solve problems in algebra, and these lessons cover all of the basic types of problems (coin, mixture, finance, etc.) found in an algebra course. Lesson 20 explains how to use algebra when solving problems in geometry. Lesson 21 explains some other types of problem-solving strategies. These strategies can be used in lieu of equations and can help in checking problems when equations are not appropriate. Because of the increasing popularity of the topics of probability and statistics, Lessons 22 and 23 cover some of the basic types of problems found in these areas. This book also contains six "Refreshers." These are intended to provide a review of topics needed to solve the word problems that follow them. They are not intended to teach the topics from scratch. You should refer to appropriate textbooks if you need additional help with the refresher topics.

This book can be used either as a self-study book or as a supplement to your textbook. You can select the lessons that are appropriate for your needs.

Best wishes on your success.

Acknowledgments

I would like to thank my wife, Betty Claire, for helping me with this project, and I wish to express my gratitude to my editor Judy Bass and to Carrie Green for their assistance in the publication of this book.

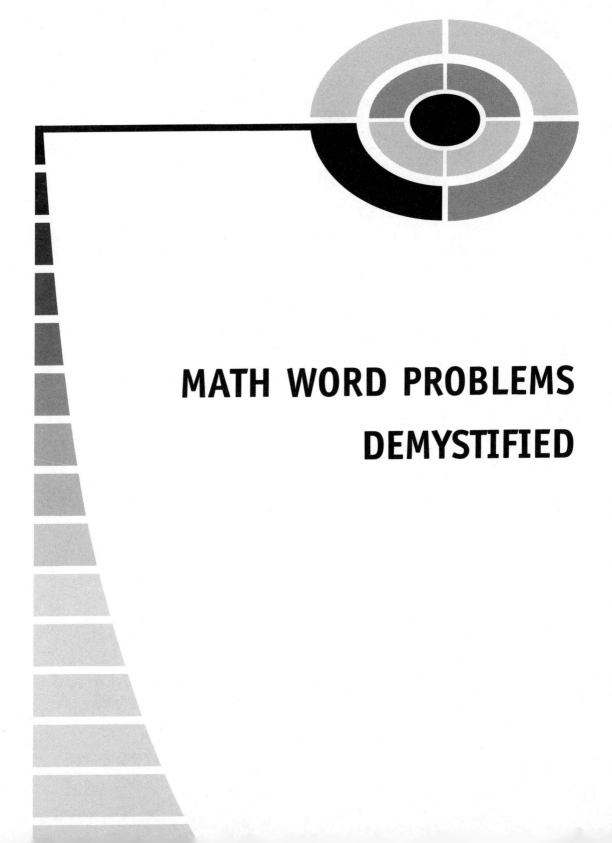

MATH WORD PROBLEMS
DEMYSTIFIED

Introduction to Solving Word Problems

In every area of mathematics, you will encounter "word" problems. Some students are very good at solving word problems while others are not. When teaching word problems in prealgebra and algebra, I often hear "I don't know where to begin," or "I have never been able to solve word problems." A great deal has been written about solving word problems. A Hungarian mathematician, George Polya, did much in the area of problem solving. His book, entitled *How To Solve It*, has been translated into at least 17 languages, and it explains the basic steps of problem solving. These steps are explained next.

Step 1. **Understand the problem.** First read the problem carefully several times. Underline or write down any information given in the

problem. Next, decide what you are being asked to find. This step is called the **goal**.

Step 2.   **Select a strategy to solve the problem.** There are many ways to solve word problems. You may be able to use one of the basic operations such as addition, subtraction, multiplication, or division. You may be able to use an equation or formula. You may even be able to solve a given problem by trial or error. This step will be called **strategy**.

Step 3.   **Carry out the strategy.** Perform the operation, solve the equation, etc., and get the solution. If one strategy doesn't work, try another one. This step will be called **implementation**.

Step 4.   **Evaluate the answer.** This means to check your answer if possible. Another way to evaluate your answer is to see if it is reasonable. Finally, you can use **estimation** as a way to check your answer. This step will be called **evaluation**.

When you think about the four steps, they apply to many situations that you may encounter in life. For example, suppose that you play basketball. The **goal** is to get the basketball into the hoop. The **strategy** is to select a way to make a basket. You can use any one of several methods, such as a jump shot, a layup, a one-handed push shot, or a slam-dunk. The strategy that you use will depend on the situation. After you decide on the type of shot to try, you **implement** the shot. Finally, you **evaluate** the action. Did you make the basket? Good for you! Did you miss it? What went wrong? Can you improve on the next shot?

Now let's see how this procedure applies to a mathematical problem.

EXAMPLE: Find the next two numbers in the sequence

$$10 \quad 8 \quad 11 \quad 9 \quad 12 \quad 10 \quad 13 \quad \underline{\hspace{1cm}} \quad \underline{\hspace{1cm}}$$

SOLUTION:

GOAL: You are asked to find the next two numbers in the sequence.

STRATEGY: Here you can use a strategy called "find a pattern." Ask yourself, "What's being done to one number to get the next number in the sequence?" In this case, to get from 10 to 8, you can subtract 2. But to get from 8 to 11, you need to add 3. So perhaps it is necessary to do two different things.

IMPLEMENTATION: Subtract 2 from 13 to get 11. Add 3 to 11 to get 14. Hence, the next two numbers should be 11 and 14.

EVALUATION: In order to check the answers, you need to see if the "subtract 2, add 3" solution works for all the numbers in the sequence, so start with 10.

$$10 - 2 = 8$$

$$8 + 3 = 11$$

$$11 - 2 = 9$$

$$9 + 3 = 12$$

$$12 - 2 = 10$$

$$10 + 3 = 13$$

$$13 - 2 = 11$$

$$11 + 3 = 14$$

Voila! You have found the solution! Now let's try another one.

EXAMPLE: Find the next two numbers in the sequence

$$1 \quad 2 \quad 4 \quad 7 \quad 11 \quad 16 \quad 22 \quad 29 \quad \underline{\hspace{1cm}} \quad \underline{\hspace{1cm}}$$

SOLUTION:

GOAL: You are asked to find the next two numbers in the sequence.

STRATEGY: Again we will use "find a pattern." Now ask yourself, "What is being done to the first number to get the second one?" Here we are adding 1. Does adding one to the second number 2 give us the third number 4? No. You must add 2 to the second number to get the third number 4. How do we get from the third number to the fourth number? Add 3. Let's apply the strategy.

IMPLEMENTATION:

$$1 + 1 = 2$$

$$2 + 2 = 4$$

$$4 + 3 = 7$$

$$7 + 4 = 11$$

$$11 + 5 = 16$$

$$16 + 6 = 22$$

$$22 + 7 = 29$$

$$29 + 8 = 37$$

$$37 + 9 = 46$$

Hence, the next two numbers in the sequence are 37 and 46.

EVALUATION: Since the pattern works for the first eight numbers in the sequence, we can extend it to the next two numbers, which then makes the answers correct.

EXAMPLE: Find the next two letters in the sequence

Z   B   Y   D   X   F   W   H   ____   ____

SOLUTION:

GOAL: You are asked to find the next two letters in the sequence.

STRATEGY: Again, you can use the "find a pattern" strategy. Notice that the sequence starts with the last letter of the alphabet Z and then goes to the second letter B, then back to the next to the last letter Y, and so on. So it looks like there are two sequences.

IMPLEMENTATION: The first sequence is Z   Y   X   W   V, and the second sequence is B   D   F   H   J. Hence, the next two letters are V and J.

EVALUATION: Putting the two sequences together, you get
Z B Y D X F W H V J.

Now, you can try a few to see if you understand the problem-solving procedure. Be sure to use all four steps.

Try These

Find the next two numbers or letters in each sequence.

1. 5   15   45   135   405   ____   ____
2. 3   6   7   14   15   30   31   _62_   _63_
3. 128   64   32   16   8   ____   ____
4. 4   9   7   12   10   15   13   ____   ____
5. 1   A   3   C   5   E   7   G   ____   ____

SOLUTIONS:

1. 1215 and 3645. The next number is 3 times the previous number.

2. 62 and 63. Multiply by 2. Add 1. Repeat.

3. 4 and 2. Divide the preceding number by 2 to get the next number.

4. 18 and 16. Add 5. Subtract 2. Repeat.

5. 9 and I. Use the odd numbers 1, 3, 5, etc., and every other letter of the alphabet, A, C, E, G, etc.

Well, how did you do? You have just had an introduction to systematic problem solving. The remainder of this book is divided into three parts. Part I explains how to solve problems in arithmetic and prealgebra. Part II explains how to solve problems in introductory and intermediate algebra and geometry. Part III explains how to solve problems using some general problem-solving strategies such as "Draw a picture," "Work backwards," etc., and how to solve problems in probability and statistics. After successfully completing this book, you will be well along the way to becoming a competent word-problem solver.

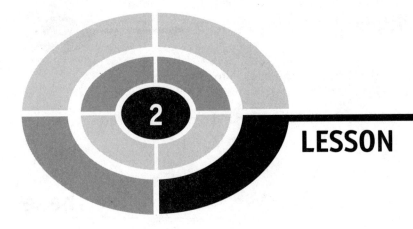

Solving Word Problems Using Whole Numbers

Most word problems in arithmetic and prealgebra can be solved by using one or more of the basic **operations**. The basic operations are addition, subtraction, multiplication, and division. Sometimes students have a problem deciding which operation to use. The correct operation can be determined by the words in the problem.

Use **addition** when you are being asked to find

the total,
the sum,
how many in all,
how many altogether, etc.,

and all the items in the problem are the same type.

EXAMPLE: In a conference center, the Mountain View Room can seat 78 people, the Lake View Room can seat 32 people, and the Trail View Room can seat 46 people. Find the total number of people that can be seated at any one time.

SOLUTION:

GOAL: You are being asked to find the total number of people that can be seated.

STRATEGY: Use addition since you need to find a total and all the items in the problem are the same (i.e., people).

IMPLEMENTATION: $78 + 32 + 46 = 156$.

EVALUATION: The conference center can seat 156 people. This can be checked by **estimation**. Round each value and then find the sum: $80 + 30 + 45 = 155$. Since the estimated sum is close to the actual sum, you can conclude that the answer is probably correct. (*Note:* When using estimation, you cannot be 100% sure your answer is correct since you have used rounded numbers.)

Use **subtraction** when you are asked to find

how much more,
how much less,
how much larger,
how much smaller,
how many more,
how many fewer,
the difference,
the balance,
how much is left,
how far above,
how far below,
how much further, etc.,

and all the items in the problems are the same type.

EXAMPLE: If Lake Erie is 241 miles long and Lake Huron is 206 miles long, how much longer is Lake Erie than Lake Huron?

SOLUTION:

GOAL: You are being asked to find how much longer Lake Erie is than Lake Huron.

STRATEGY: Since you are being asked "how much longer" and both items are the same (miles), you use subtraction.

IMPLEMENTATION: 241 miles − 206 miles = 35 miles. Hence, Lake Erie is 35 miles longer than Lake Huron.

EVALUATION: You can check the solution by adding: 206 + 35 = 241.

Use **multiplication** when you are being asked to find

the product,
the total,
how many in all,
how many altogether, etc.,

and you have so many groups of individual items.

EXAMPLE: Find the total number of seats in an auditorium if there are 22 rows with 36 seats in a row.

SOLUTION:

GOAL: You are being asked to find the **total** number of seats in the auditorium.

STRATEGY: Use multiplication because you are asked to find a total and you have 22 rows (groups) of seats with 36 (individual items) in a row.

IMPLEMENTATION: 36 × 22 = 792 seats. Hence, the auditorium has 792 seats.

EVALUATION: You can check your answer by estimation: 40 × 20 = 800. Since 792 is close to 800, your answer is probably correct.

Use **division** when you are given the total number of items and a number of groups and need to find how many items in each group or when you are given the total number of items and the number of items in each group and need to find how many groups.

EXAMPLE: The shipping department of a business needs to ship 496 calculators. If they are packed 8 per box, how many boxes will be needed?

SOLUTION:

GOAL: You are being asked to find how many boxes are needed.

STRATEGY: Here you are given the **total** number of calculators, 496, and need to pack 8 items in each box. You are asked to find how many boxes (groups) are needed. In this case, use division.

IMPLEMENTATION: 496 ÷ 8 = 62 boxes. Hence, you will need 62 boxes.

EVALUATION: Check: 62 boxes × 8 calculators per box = 496 calculators.

Now you can see how to decide what operation to use to solve arithmetic or preaglebra problems using whole numbers.

Try These

1. The Rolling Stones tour grossed $121,200,000 in 1994 while Pink Floyd grossed $103,500,000 in 1994. How much more money did the Rolling Stones make than Pink Floyd in 1994?

2. In New Jersey, the federal government owns 129,791 acres of land. In Texas, the federal government owns 2,307,171 acres of land, and in Maryland, the federal government owns 166,213 acres of land. Find the total amount of land owned by the federal government in the three states.

3. If 5 DVD players cost $645, find the cost of one of them.

4. If you can burn 50 calories by swimming for 1 minute, how many calories can be burned when you swim for 15 minutes?

5. If a person traveled 3588 miles and used 156 gallons of gasoline, find the miles per gallon.

6. The highest point in Alaska on top of Mt. McKinley is 20,320 feet. The highest point in Florida in Walton County is 345 feet. How much higher is the point in Alaska than the point in Florida?

7. The length of Lake Ontario is 193 miles, the length of Lake Erie is 241 miles, and the length of Lake Huron is 206 miles. How far does a person travel if he navigates all three lakes?

8. If a person needs 2500 sheets of paper, how many 500-page reams does she have to buy?

9. If you borrow $1248 from your brother and pay it back in 8 equal monthly payments, how much would you pay each month? (Your brother isn't charging you interest.)

10. If Keisha bought 9 picture frames at $19 each, find the total cost of the frames.

SOLUTIONS:

1. $121,200,000 - $103,500,000 = $17,700,000

2. $129,791 + 2,307,171 + 166,213 = 2,603,175$ acres

3. $645 \div 5 = $129

4. $15 \times 50 = 750$ calories

5. $3588 \div 156 = 23$ miles per gallon

6. $20,320 - 345 = 19,975$ feet

7. $193 + 241 + 206 = 640$ miles

8. $2500 \div 500 = 5$ reams

9. $1248 \div 8 = $156

10. $19 \times 9 = $171

Decimals

To add or subtract decimals, place the numbers in a vertical column and line up the decimal points. Add or subtract as usual and place the decimal point in the answer directly below the decimal points in the problem.

EXAMPLE: Find the sum: 32.6 + 231.58 + 6.324.

SOLUTION:

$$
\begin{array}{r}
32.600 \\
231.580 \\
+\ 6.324 \\
\hline
270.504
\end{array}
$$
Zeros can be written
to keep the columns
in line

EXAMPLE: Subtract 15.8 − 5.326.

SOLUTION:

$$
\begin{array}{r}
15.800 \\
-5.326 \\
\hline
10.474
\end{array}
$$

To multiply two decimals, multiply the numbers as is usually done. Count the number of digits to the right of the decimal points in the problem and then have the same number of digits to the right of the decimal point in the answer.

EXAMPLE: Multiply 28.62 × 3.7.

SOLUTION:

```
      28.62    You need
   ×   3.7     3 decimal
      20034    places in the
       8586    answer
     105.894
```

To divide two decimals when there is no decimal point in the divisor (the number outside the division box), place the decimal point in the answer directly above the decimal point in the dividend (the number under the division box). Divide as usual.

EXAMPLE: Divide 2305.1 ÷ 37.

SOLUTION:

```
          62.3
    37 ) 2305.1
         222
          85
          74
         111
         111
```

To divide two decimals when there is a decimal point in the divisor, move the point to the end of the number in the divisor, and then move the point the same number of places in the dividend. Place the decimal point in the answer directly above the decimal point in the dividend. Divide as usual.

EXAMPLE: Divide 30.651 ÷ 6.01.

SOLUTION:

```
                      5.1
  6.01 ) 30.651   601 ) 3065.1    Move the points
                       3005        two places to the
                        601        right
                        601
```

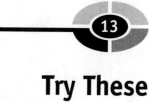

Try These

1.   Add $42.6 + 37.23 + 3.215$.
2.   Subtract $87.6 - 51.35$.
3.   Multiply 625.1×2.7.
4.   Divide $276.3 \div 45$.
5.   Divide $4.864 \div 3.2$.

SOLUTIONS:

1.
$$
\begin{array}{r}
42.600 \\
37.230 \\
+ \ \ 3.215 \\
\hline
83.045
\end{array}
$$

2.
$$
\begin{array}{r}
87.60 \\
-51.35 \\
\hline
36.25
\end{array}
$$

3.
$$
\begin{array}{r}
625.1 \\
\times \ \ \ 2.7 \\
\hline
43757 \\
12502 \ \ \\
\hline
1687.77
\end{array}
$$

4.
$$
\begin{array}{r}
6.14 \\
45\overline{)276.30} \\
270 \ \ \ \ \ \\
\hline
63 \ \ \ \\
45 \ \ \ \\
\hline
180 \ \\
180 \ \\
\hline
\end{array}
$$

A zero was written to complete the problem

5.
$$
\begin{array}{r}
1.52 \\
32\overline{)48.64} \\
32 \ \ \ \ \\
\hline
166 \ \\
160 \ \\
\hline
64 \\
64 \\
\hline
\end{array}
$$

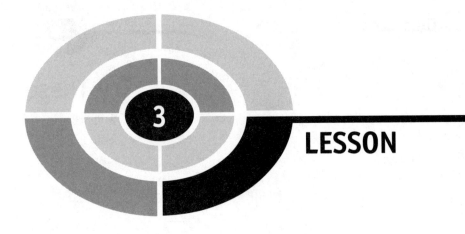

LESSON

3

Solving Word Problems Using Decimals

If you need to review decimals, complete Refresher I.

In order to solve word problems involving decimals, use the same strategies that you used in Lesson 2.

EXAMPLE: If a lawnmower uses 0.6 of a gallon of gasoline per hour, how many gallons of gasoline will be used if it takes 2.6 hours to cut a lawn?

SOLUTION:

GOAL: You are being asked to find the total number of gallons of gasoline used.

STRATEGY: Since you need to find a total and you are given two different items (gallons and hours), you multiply.

IMPLEMENTATION: $0.6 \times 2.6 = 1.56$ gallons.

EVALUATION: You can check your answer using estimation. You use about one half of a gallon per hour. In two hours you would use about one gallon and another half of a gallon for the last half hour, so you would use approximately one and one half gallons. This is close to 1.56 gallons since one and one half is 1.5.

EXAMPLE: Before Harry left on a trip, his odometer read 46351.6. After the trip, the odometer reading was 47172.9. How long was the trip?

SOLUTION:

GOAL: You are being asked to find the distance the automobile traveled.

STRATEGY: In order to find the distance, you need to subtract the two odometer readings.

IMPLEMENTATION: $47,172.9 - 46,351.6 = 821.3$ miles.

EVALUATION: Estimate the answer by rounding 47,172.9 to 47,000 and 46,351.6 to 46,000; then subtract $47,000 - 46,000 = 1000$ miles. Since 821.3 is close to 1000, the answer is probably correct.

   Sometimes, a word problem requires two or more steps. In this situation, you still follow the suggestions given in Lesson 2 to determine the operations.

EXAMPLE: Find the total cost of 6 electric keyboards at $149.97 each and 3 digital drums at $69.97 each.

SOLUTION:

GOAL: You are being asked to find the total cost of 2 different items: 6 of one item and 3 of another item.

STRATEGY: Use multiplication to find the total cost of the keyboards and the digital drums, and then add the answers.

IMPLEMENTATION: The cost of the keyboards is $6 \times \$149.97 = \$899.82$. The cost of the digital drums is $3 \times \$69.97 = \$209.91$. Add the two answers: $\$899.82 + \$209.91 = \$1109.73$. Hence, the total cost of 6 keyboards and 3 digital drums is $\$1109.73$.

EVALUATION: Estimate the answer: Keyboards: $6 \times \$150 = \$900$, Digital drums: $3 \times \$70 = \$210$, Total cost: $\$900 + \$210 = \$1110$. The estimated cost of $\$1110$ is close to the computed actual cost of $\$1109.73$; therefore, the answer is probably correct.

Try These

1. * Find the cost of 6 wristwatches at $29.95 each.

2. Yesterday the high temperature was 73.5 degrees, and today the high temperature was 68.8 degrees. How much warmer was it yesterday?

3. If the total cost of four CDs is $59.80, find the cost of each one.

4. Kamel made six deliveries today. The distances he drove were 6.32 miles, 4.81 miles, 15.3 miles, 3.72 miles, 5.1 miles, and 9.63 miles. Find the total miles he drove.

5. A person mixed 26.3 ounces of water with 22.4 ounces of alcohol. Find the total number of ounces of solution.

6. Find the total cost of 6 pairs of boots at $49.95 each and 5 pairs of gloves at $14.98 each.

7. ✏ Beth bought 2 pairs of sunglasses at $19.95 each. If she paid for them with a $50.00 bill, how much change did she receive?

8. The weight of water is 62.5 pounds per cubic foot. Find the total weight of a container if it holds 6 cubic feet of water and the empty container weighs 30.6 pounds.

9. A taxi driver charges $10.00 plus $4.75 per mile. Find the total cost of a 7-mile trip.

10. Wendel earns $6.50 per hour and $9.75 for each hour over 40 hours per week. Find his earnings if he worked 44 hours last week.

SOLUTIONS:

1. $6 \times \$29.95 = \$179.70.$

2. $73.5° - 68.8° = 4.7°.$

3. $\$59.80 \div 4 = \$14.95.$

4. $6.32 + 4.81 + 15.3 + 3.72 + 5.1 + 9.63 = 44.88$ miles.

5. $26.3 + 22.4 = 48.7.$

6. $6 \times \$49.95 + 5 \times \$14.98 = \$299.70 + \$74.90 = \$374.60.$

7. $2 \times \$19.95 = \$39.90; \$50.00 - \$39.90 = \$10.10.$

8. $6 \times 62.5 = 375; 375 + 30.6 = 405.6$ pounds.

9. $7 \times \$4.75 = \$33.25; \$33.25 + \$10.00 = \$43.25.$

10. $40 \times \$6.50 = \$260.00; 4 \times \$9.75 = \$39.00; \$260.00 + \$39.00 = \$299.00.$

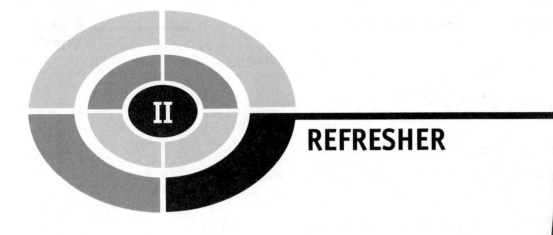

Fractions

In a fraction, the top number is called the **numerator** and the bottom number is called the **denominator**.

To reduce a fraction to lowest terms, divide the numerator and denominator by the largest number that divides evenly into both.

EXAMPLE: Reduce $\dfrac{24}{32}$.

SOLUTION:

$$\frac{24}{32} = \frac{24 \div 8}{32 \div 8} = \frac{3}{4}$$

To change a fraction to higher terms, divide the smaller denominator into the larger denominator, and then multiply the smaller numerator by that number.

EXAMPLE: Change $\dfrac{5}{6}$ to 24ths.

SOLUTION:

Divide $24 \div 6 = 4$ and multiply $5 \times 4 = 20$. Hence, $\dfrac{5}{6} = \dfrac{20}{24}$.

This can be written as $\dfrac{5}{6} = \dfrac{5 \times 4}{6 \times 4} = \dfrac{20}{24}$.

An **improper fraction** is a fraction whose numerator is greater than or equal to its denominator. For example, $\frac{20}{3}$, $\frac{6}{5}$, and $\frac{3}{3}$ are improper fractions. A **mixed number** is a whole number and a fraction; $8\frac{1}{3}$, $2\frac{1}{4}$, and $3\frac{5}{6}$ are mixed numbers.

To change an improper fraction to a mixed number, divide the numerator by the denominator and write the remainder as the numerator of a fraction whose denominator is the divisor. Reduce the fraction if possible.

EXAMPLE: Change $\frac{21}{6}$ to a mixed number.

SOLUTION:

$$\begin{array}{r} 3 \\ 6\overline{)21} \\ \underline{18} \\ 3 \end{array} \qquad \frac{21}{6} = 3\frac{3}{6} = 3\frac{1}{2}$$

To change a mixed number to an improper fraction, multiply the denominator of the fraction by the whole number and add the numerator. This will be the numerator of the improper fraction. Use the same number for the denominator of the improper fraction as the number in the denominator of the fraction in the mixed number.

EXAMPLE: Change $5\frac{2}{3}$ to an improper fraction.

SOLUTION:

$$5\frac{2}{3} = \frac{3 \cdot 5 + 2}{3} = \frac{17}{3}$$

In order to add or subtract fractions, you need to find the **lowest common denominator** of the fractions. The lowest common denominator (LCD) of the fractions is the smallest number that can be divided evenly by all the denominator numbers. For example, the LCD of $\frac{1}{6}$, $\frac{2}{3}$, and $\frac{7}{9}$ is 18, since 18 can be divided evenly by 3, 6, and 9. There are several mathematical methods for finding the LCD; however, we will use the guess method. That is, just look at the denominators and figure out the LCD. If needed, you can look at an arithmetic or prealgebra book for a mathematical method to find the LCD.

To add or subtract fractions:

1. Find the LCD.
2. Change the fractions to higher terms.
3. Add or subtract the numerators. Use the LCD.
4. Reduce or simplify the answer if necessary.

EXAMPLE: Add $\dfrac{3}{4} + \dfrac{5}{6} + \dfrac{2}{3}$.

SOLUTION:

Use 12 as the LCD.

$$\frac{3}{4} = \frac{9}{12}$$

$$\frac{5}{6} = \frac{10}{12}$$

$$+\frac{2}{3} = \frac{8}{12}$$

$$\frac{27}{12} = 2\frac{3}{12} = 2\frac{1}{4}$$

EXAMPLE: Subtract $\dfrac{9}{10} - \dfrac{3}{8}$.

SOLUTION:

Use 40 as the LCD.

$$\frac{9}{10} = \frac{36}{40}$$

$$-\frac{3}{8} = \frac{15}{40}$$

$$\frac{21}{40}$$

To multiply two or more fractions, cancel if possible, multiply numerators, and then multiply denominators.

EXAMPLE: Multiply $\dfrac{3}{8} \times \dfrac{4}{9}$.

SOLUTION:

$$\frac{3}{8} \times \frac{4}{9} = \frac{\cancel{3}^{1}}{\cancel{8}_{2}} \times \frac{\cancel{4}^{1}}{\cancel{9}_{3}} = \frac{1 \times 1}{2 \times 3} = \frac{1}{6}$$

To divide two fractions, invert (turn upside down) the fraction after the $\div$ sign and multiply.

EXAMPLE: Divide $\dfrac{9}{10} \div \dfrac{3}{5}$.

SOLUTION:

$$\frac{9}{10} \div \frac{3}{5} = \frac{9}{10} \times \frac{5}{3} = \frac{\cancel{9}^{3}}{\cancel{10}_{2}} \times \frac{\cancel{5}^{1}}{\cancel{3}_{1}} = \frac{3 \times 1}{2 \times 1} = \frac{3}{2} = 1\frac{1}{2}$$

To add mixed numbers, add the fractions, add the whole numbers, and simplify the answer if necessary.

EXAMPLE: Add $1\dfrac{5}{6} + 4\dfrac{7}{8}$.

SOLUTION:

$$1\frac{5}{6} = 1\frac{20}{24}$$
$$+4\frac{7}{8} = 4\frac{21}{24}$$
$$5\frac{41}{24} = 5 + 1\frac{17}{24} = 6\frac{17}{24}$$

To subtract mixed numbers, borrow if necessary, subtract the fractions, and then subtract the whole numbers.

EXAMPLE: Subtract $6\dfrac{9}{10} - 2\dfrac{2}{3}$.

SOLUTION:

$$6\frac{9}{10} = 6\frac{27}{30}$$

$$-2\frac{2}{3} = 2\frac{20}{30}$$

$$4\frac{7}{30}$$

No borrowing is necessary here.

When borrowing is necessary, take one away from the whole number and add it to the fraction. For example,

$$8\frac{3}{5} = 8 + \frac{3}{5} = 7 + 1 + \frac{3}{5} = 7 + \frac{5}{5} + \frac{3}{5} = 7\frac{8}{5}$$

Another example:

$$11\frac{4}{9} = 11 + \frac{4}{9} = 10 + 1 + \frac{4}{9} = 10 + \frac{9}{9} + \frac{4}{9} = 10\frac{13}{9}$$

EXAMPLE: Subtract $7\frac{1}{4} - 3\frac{5}{8}$.

SOLUTION:

$$7\frac{1}{4} = 7\frac{2}{8} = 6\frac{10}{8}$$

$$-3\frac{5}{8} = 3\frac{5}{8} = 3\frac{5}{8}$$

$$3\frac{5}{8}$$

To multiply or divide mixed numbers, change the mixed numbers to improper fractions, and then multiply or divide as shown before.

EXAMPLE: Multiply $3\frac{3}{8} \times 2\frac{2}{3}$.

SOLUTION:

$$3\frac{3}{8} \times 2\frac{2}{3} = \frac{27^9}{8_1} \times \frac{8^1}{3_1} = \frac{9}{1} = 9$$

EXAMPLE: Divide $5\frac{1}{2} \div 3\frac{1}{4}$

SOLUTION:

$$5\frac{1}{2} \div 3\frac{1}{4} = \frac{11}{2} \div \frac{13}{4} = \frac{11}{2_1} \times \frac{4^2}{13} = \frac{22}{13} = 1\frac{9}{13}$$

To change a fraction to a decimal, divide the numerator by the denominator.

EXAMPLE: Change $\frac{5}{8}$ to a decimal.

SOLUTION:

$$
\begin{array}{r}
.625 \\
8\overline{)5.00} \\
48 \\
\hline
20 \\
16 \\
\hline
40 \\
40 \\
\hline
\end{array}
\qquad \frac{5}{8} = 0.625
$$

To change a decimal to a fraction, drop the decimal point and place the number over 10 if it has one decimal place, 100 if it has two decimal places, 1000 if it has three decimal places, etc. Reduce if possible.

EXAMPLE: Change 0.45 to a fraction.

SOLUTION:

$$0.45 = \frac{45}{100} = \frac{9}{20}$$

Try These

1. Reduce to lowest terms: $\dfrac{12}{30}$.

2. Add $\dfrac{2}{3} + \dfrac{3}{4} + \dfrac{1}{8}$.

3. Subtract $\dfrac{11}{12} - \dfrac{5}{9}$.

4. Multiply $\dfrac{5}{6} \times \dfrac{3}{10}$.

5. Divide $\dfrac{8}{9} \div \dfrac{2}{3}$.

6. Add $1\dfrac{1}{2} + 3\dfrac{5}{6}$.

7. Subtract $12\dfrac{1}{3} - 2\dfrac{3}{5}$.

8. Multiply $3\dfrac{3}{4} \times 2\dfrac{2}{5}$.

9. Divide $4\dfrac{1}{2} \div 1\dfrac{1}{2}$.

10. Change $\dfrac{5}{16}$ to a decimal.

11. Change 0.64 to a fraction.

SOLUTIONS:

1. $\dfrac{12}{30} = \dfrac{12 \div 6}{30 \div 6} = \dfrac{2}{5}$

2.

$$\frac{2}{3} = \frac{16}{24}$$

$$\frac{3}{4} = \frac{18}{24}$$

$$+\ \frac{1}{8} = \frac{3}{24}$$

$$\frac{37}{24} = 1\frac{13}{24}$$

3.

$$\frac{11}{12} = \frac{33}{36}$$

$$-\frac{5}{9} = \frac{20}{36}$$

$$\frac{13}{36}$$

4. $\dfrac{5}{6} \times \dfrac{3}{10} = \dfrac{\cancel{5}^{1}}{\cancel{6}^{2}} \times \dfrac{\cancel{3}^{1}}{\cancel{10}^{2}} = \dfrac{1 \times 1}{2 \times 2} = \dfrac{1}{4}$

5. $\dfrac{8}{9} \div \dfrac{2}{3} = \dfrac{\cancel{8}^{4}}{\cancel{9}^{3}} \times \dfrac{\cancel{3}^{1}}{\cancel{2}^{1}} = \dfrac{4}{3} = 1\dfrac{1}{3}$

6.

$$1\frac{1}{2} = 1\frac{3}{6}$$

$$+3\frac{5}{6} = 3\frac{5}{6}$$

$$4\frac{8}{6} = 5\frac{2}{6} = 5\frac{1}{3}$$

7. $12\dfrac{1}{3} = 12\dfrac{5}{15} = 11\dfrac{20}{15}$

$$-2\frac{3}{5} = \ 2\frac{9}{15} = 2\frac{9}{15}$$

$$9\frac{11}{15}$$

8. $3\dfrac{3}{4} \times 2\dfrac{2}{5} = \dfrac{\cancel{15}^{3}}{\cancel{4}^{1}} \times \dfrac{\cancel{12}^{3}}{\cancel{5}^{1}} = \dfrac{9}{1} = 9$

9. $4\dfrac{1}{2} \div 1\dfrac{1}{2} = \dfrac{9}{2} \div \dfrac{3}{2} = \dfrac{\cancel{9}^{3}}{\cancel{2}^{1}} \times \dfrac{\cancel{2}^{1}}{\cancel{3}^{1}} = \dfrac{3}{1} = 3$

10.
$$
\begin{array}{r}
.3125 \\
16\overline{)5.000} \\
\underline{48} \\
20 \\
\underline{16} \\
40 \\
\underline{32} \\
80 \\
\underline{80}
\end{array}
\qquad \dfrac{5}{16} = 0.3125
$$

11. $0.64 = \dfrac{64}{100} = \dfrac{16}{25}$

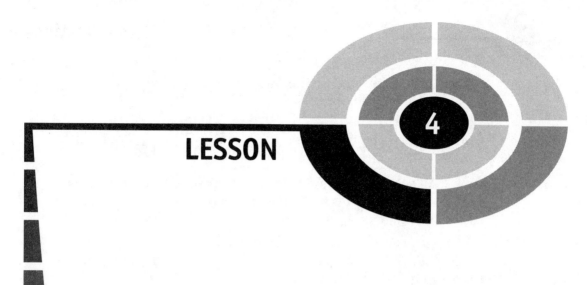

Solving Word Problems Using Fractions

If you need to review fractions, complete Refresher II.

In order to solve word problems involving fractions, use the same strategies that you used in Lesson 2.

EXAMPLE: Find the total thickness of three pieces of wood that are nailed together, if one is $1\frac{1}{4}$ in. thick, one is $\frac{7}{8}$ in. thick, and one is $2\frac{3}{16}$ in. thick.

SOLUTION:

GOAL: You are being asked to find the total thickness of three pieces of wood.

STRATEGY: Since you need to find a total and all items are in the same units (inches), use addition.

IMPLEMENTATION: $1\dfrac{1}{4} + \dfrac{7}{8} + 2\dfrac{3}{16} = 1\dfrac{4}{16} + \dfrac{14}{16} + 2\dfrac{3}{16} = 3\dfrac{21}{16} = 4\dfrac{5}{16}$

EVALUATION: You can estimate the answer since $1\frac{1}{4}$ in. is about 1 in., $\frac{7}{8}$ in. is about 1 in., and $2\frac{3}{16}$ in. is about 2 in. Hence, 1 in. + 1 in. + 2 in. = 4 in. Since 4 in. is close to $4\frac{5}{16}$ in., your answer is probably correct.

EXAMPLE: How many pieces of ribbon $5\frac{1}{2}$ inches long can be cut from a piece of ribbon $38\frac{1}{2}$ inches in length?

SOLUTION:

GOAL: You are being asked to find the number of pieces of ribbon that can be cut from a piece $38\frac{1}{2}$ inches long.

STRATEGY: Since you are given a total ($38\frac{1}{2}$ in.) and need to find how many pieces ($5\frac{1}{2}$ in.) are in the total, you can use division.

IMPLEMENTATION: $38\dfrac{1}{2} \div 5\dfrac{1}{2} = \dfrac{77}{2} \div \dfrac{11}{2} = \dfrac{\cancel{77}^{7}}{\cancel{2}^{1}} \times \dfrac{\cancel{2}^{1}}{\cancel{11}^{1}} = 7$

Hence, you can cut 7 pieces.

EVALUATION: You can check by multiplying $7 \times 5\frac{1}{2} = 38\frac{1}{2}$

EXAMPLE: A person has hiked $1\frac{3}{4}$ miles by noon. If his destination is $5\frac{7}{8}$ miles away from his starting point, how much further does he have to go?

SOLUTION:

GOAL: You are being asked to find how much further the person has to go.

STRATEGY: Since you need to find how much further, you use subtraction.

IMPLEMENTATION: $5\dfrac{7}{8} - 1\dfrac{3}{4} = 5\dfrac{7}{8} - 1\dfrac{6}{8} = 4\dfrac{1}{8}$ miles.

EVALUATION: You can check the solution by adding
$1\frac{3}{4} + 4\frac{1}{8} = 1\frac{6}{8} + 4\frac{1}{8} = 5\frac{7}{8}$

29 —

Try These

1. How many pieces of pipe $3\frac{5}{8}$ inches long can be cut from a piece 29 inches long?

2. If the temperature is $62\frac{1}{2}°$ and drops $15\frac{3}{4}°$, what is the present temperature?

3. Find the total length of 5 pieces of wood if each piece is $2\frac{1}{4}$ inches long.

4. Find the total amount of wheat in 3 containers each holding $2\frac{1}{6}$ bushels, $1\frac{3}{8}$ bushels, and $4\frac{1}{3}$ bushels of wheat respectively.

5. A recipe that serves 4 people calls for $3\frac{1}{2}$ cups of flour. How much flour is needed if the chef wants to serve 12 people?

6. A candy store owner mixed $2\frac{7}{8}$ pounds of caramels with $3\frac{5}{6}$ pounds of chocolate. How much candy did he have in all?

7. If you need to drive $8\frac{2}{3}$ miles to pick up your friend and then drive $5\frac{3}{5}$ miles to a mall, how far did you drive to and from your home?

8. Peter is $48\frac{3}{4}$ inches tall. His sister is $52\frac{1}{8}$ inches tall. How much taller is Peter's sister than Peter?

9. A box of cookies weighs $14\frac{3}{4}$ ounces. How much does a carton containing 12 boxes weigh?

10. How much wood would be needed to frame a $8\frac{1}{2}$ inch by 11 inch picture?

SOLUTIONS:

1. $29 \div 3\frac{5}{8} = \frac{29}{1} \div \frac{29}{8} = \frac{29}{1} \times \frac{8}{29} = \frac{8}{1} = 8$

2. $62\frac{1}{2} - 15\frac{3}{4} = 62\frac{2}{4} - 15\frac{3}{4} = 61\frac{6}{4} - 15\frac{3}{4} = 46\frac{3}{4}$

3. $5 \times 2\frac{1}{4} = \frac{5}{1} \times \frac{9}{4} = \frac{45}{4} = 11\frac{1}{4}$ inches

4. $2\frac{1}{6} + 1\frac{3}{8} + 4\frac{1}{3} = 2\frac{4}{24} + 1\frac{9}{24} + 4\frac{8}{24} = 7\frac{21}{24} = 7\frac{7}{8}$ bushels

5. $12 \div 4 = 3; \; 3 \times 3\frac{1}{2} = \frac{3}{1} \times \frac{7}{2} = \frac{21}{2} = 10\frac{1}{2}$ cups

6. $2\frac{7}{8} + 3\frac{5}{6} = 2\frac{21}{24} + 3\frac{20}{24} = 5\frac{41}{24} = 6\frac{17}{24}$ pounds

7. $8\frac{2}{3} + 5\frac{3}{5} = 8\frac{10}{15} + 5\frac{9}{15} = 13\frac{19}{15} = 14\frac{4}{15};$

    $14\frac{4}{15} \times 2 = \frac{214}{15} \times \frac{2}{1} = \frac{428}{15} = 28\frac{8}{15}$ miles

8. $52\frac{1}{8} - 48\frac{3}{4} = 52\frac{1}{8} - 48\frac{6}{8} = 51\frac{9}{8} - 48\frac{6}{8} = 3\frac{3}{8}$ inches

9. $14\frac{3}{4} \times 12 = \frac{59}{4} \times \frac{12}{1} = 177$ ounces

10. $8\frac{1}{2} + 11 + 8\frac{1}{2} + 11 = 39$ inches

Quiz 1

1. If a person earns $54,000 a year, what is the person's monthly salary?
   (a) $4000
   (b) $4200
   (c) $4500
   (d) $4800

2. The Arkansas River is 1459 miles long and the Delaware River is 390 miles long. How much longer is the Arkansas River than the Delaware River?
   (a) 1849 miles
   (b) 1069 miles
   (c) 1129 miles
   (d) 1589 miles

3. Find the cost of 24 feet of chain if it sells for $1.29 per foot.
   (a) $32.25
   (b) $31.20
   (c) $32.40
   (d) $30.96

4. Pennsylvania has 501 public school districts, Ohio has 661 school districts, and New York has 705 school districts. Find the total number of school districts in all three states.
   (a) 1867
   (b) 1206
   (c) 1366
   (d) 1162

5. A professor said $\frac{4}{5}$ of his students passed his final examination. If 60 students took the exam, how many passed?
   (a) 42
   (b) 56
   (c) 48
   (d) 75

6. A carpenter made 5 shelves that were $3\frac{7}{8}$ feet long and 4 shelves that were $4\frac{1}{4}$ feet long. How much lumber did he use?
   (a) $36\frac{3}{8}$ inches
   (b) $73\frac{1}{8}$ inches
   (c) $36\frac{3}{4}$ inches
   (d) $72\frac{1}{2}$ inches

7. A person made the following purchases: $42.50, $39.98, $87.49, and $16.20. How much did this person spend in all?
   (a) $184.27
   (b) $188.77
   (c) $186.17
   (d) $192.37

8. A person purchased a refrigerator for $60 down and 8 monthly payments of $56.60. Find the total cost of the refrigerator.
   (a) $536.60
   (b) $512.80
   (c) $452.80
   (d) $641.60

9. How much ribbon is needed to make 23 Christmas tree ornaments if each ornament requires $4\frac{5}{6}$ inches of ribbon?
    (a) $111\frac{1}{6}$ inches
    (b) $4\frac{22}{29}$ inches
    (c) $27\frac{4}{6}$ inches
    (d) $92\frac{5}{6}$ inches

10. How many hours can a diesel engine run on a 600-gallon tank of fuel if it uses $\frac{7}{8}$ of a gallon per hour?
    (a) 525 hours
    (b) 875 hours
    (c) $650\frac{3}{7}$ hours
    (d) $685\frac{5}{7}$ hours

Percents

Percent means hundredths or part of a hundred. For example, 37% means 0.37 or $\frac{37}{100}$. You can think of 37% as a square being divided into 100 equal parts and 37% is 37 equal parts out of 100 equal parts.

To change a percent to a decimal, drop the % sign and move the decimal point two places to the left. The decimal point in 37% is between the 7 and the % sign. It is not written.

EXAMPLE: Write each percent as a decimal:

   (a)   54%
   (b)   6%
   (c)   235%
   (d)   42.8%

SOLUTION:

   (a)   54% = 0.54
   (b)   6% = 0.06
   (c)   235% = 2.35
   (d)   42.8% = 0.428

To change a decimal to a percent, move the decimal two places to the right and affix the percent sign.

EXAMPLE: Change each decimal to a percent:

(a) 0.83
(b) 0.05
(c) 4.32
(d) 0.137

SOLUTION:

(a) $0.83 = 83\%$
(b) $0.05 = 5\%$
(c) $4.32 = 432\%$
(d) $0.137 = 13.7\%$

To change a percent to a fraction, drop the percent sign and place the number in the numerator of a fraction whose denominator is 100. Reduce or simplify if necessary.

EXAMPLE: Change each percent to a fraction:

(a) 40%
(b) 66%
(c) 125%
(d) 8%

SOLUTION:

(a) $40\% = \dfrac{40}{100} = \dfrac{2}{5}$

(b) $66\% = \dfrac{66}{100} = \dfrac{33}{50}$

(c) $125\% = \dfrac{125}{100} = \dfrac{5}{4} = 1\dfrac{1}{4}$

(d) $8\% = \dfrac{8}{100} = \dfrac{2}{25}$

To change a fraction to a percent, change the fraction to a decimal and then change the decimal to a percent.

EXAMPLE: Change each fraction or mixed number to a percent:

(a) $\dfrac{3}{5}$

(b) $\dfrac{9}{10}$

(c) $\dfrac{5}{8}$

(d) $1\dfrac{1}{4}$

SOLUTION:

(a)
$$5\overline{)3.0} \quad \dfrac{.6}{}$$
$$\dfrac{30}{0}$$
$$\dfrac{3}{5} = 0.6 = 60\%$$

(b)
$$10\overline{)9.0} \quad \dfrac{.9}{}$$
$$\dfrac{90}{0}$$
$$\dfrac{9}{10} = 0.9 = 90\%$$

(c)
$$8\overline{)5.00} \quad \dfrac{.625}{}$$
$$\dfrac{48}{20}$$
$$\dfrac{16}{40}$$
$$\dfrac{40}{0}$$
$$\dfrac{5}{8} = 0.625 = 62.5\%$$

(d) $1\dfrac{1}{4} = \dfrac{5}{4}$
$$4\overline{)5.00} \quad \dfrac{1.25}{}$$
$$\dfrac{4}{10}$$
$$\dfrac{8}{20}$$
$$\underline{20}$$
$$1\dfrac{1}{4} = 1.25 = 125\%$$

A percent word problem has three numbers: the whole, total, or base (*B*); the part (*P*); and the rate or percent (*R*). Suppose that in a class of 30 students, there are 6 absent. Now the whole or total is 30 and the part is 6. The rate or percent of students who were absent is $\frac{6}{30} = \frac{1}{5} = 20\%$.

In a percent problem, you will be given two of the three numbers and will be asked to find the third number. Percent problems can be solved by using a percent circle. The circle is shown in Figure RIII-1.

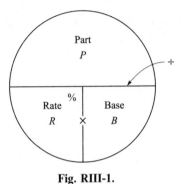

Fig. RIII-1.

In the top portion of the circle, write the word part (P). In the lower left portion of the circle, write the word rate (R), and in the lower right portion, write the word base (B). Put a multiplication sign between the two lower portions and a division sign between the top and bottom portions.

If you are asked to find the part (P), place the rate (R) in the lower left portion of the circle and the base (B) in the lower right portion. The circle tells you to use the formula $P = R \cdot B$ and multiply.

If you are asked to find the rate (R), place the part (P) in the top portion of the circle and the base (B) in the lower right portion. The circle tells you to use the formula $R = \frac{P}{B}$ and divide. The answer will be in decimal form and needs to be changed to a percent.

If you are asked to find the base (B), place the part (P) in the top portion and the rate (R) in the bottom left portion. The circle tells you to use the formula $B = \frac{P}{R}$ and divide. See Figure RIII-2. **Be sure to change the percent to a decimal or fraction before multiplying or dividing.**

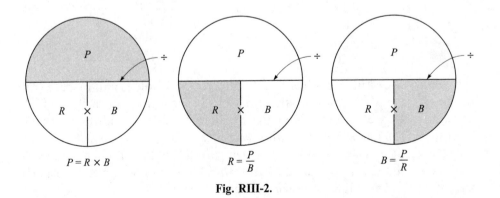

Fig. RIII-2.

TYPE I: FINDING THE PART

EXAMPLE: Find 30% of 18.

SOLUTION: Since 30% is the rate, place it in the lower left portion of the circle, and since 18 is the base, place it in the lower right portion of the circle and then multiply. See Figure RIII-3.

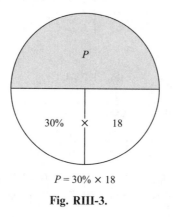

$$P = 30\% \times 18$$

Fig. RIII-3.

$$P = R \cdot P$$
$$= 30\% \cdot 18$$
$$= 0.30 \times 18$$
$$= 5.4$$

Note: The number after the word "of" is the base.

TYPE 2: FINDING THE RATE

EXAMPLE: 12 is what percent of 20?

SOLUTION: Since 12 is the part, place it in the top portion of the circle, and since 20 is the base, place it in the lower right portion of the circle, and then divide. See Figure RIII-4.

$$R = \frac{P}{B}$$
$$= \frac{12}{20}$$
$$= 0.60 = 60\%$$

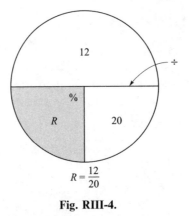

$$R = \frac{12}{20}$$

Fig. RIII-4.

TYPE 3: FINDING THE BASE

EXAMPLE: 32 is 80% of what number?

SOLUTION: Since 32 is the part, place it in the top portion of the circle, and since 80% is the rate, place it in the lower left portion of the circle and then divide. See Figure RIII-5.

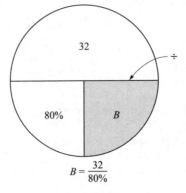

$$B = \frac{32}{80\%}$$

Fig. RIII-5.

$$B = \frac{P}{R}$$

$$= \frac{32}{80\%}$$

$$= \frac{32}{0.80}$$

$$= 40$$

Try These

1. Find 65% of 50.

2. What percent of 40 is 25?

3. 18 is 30% of what number?

4. What percent of 25 is 15?

5. 15 is 75% of what number?

6. Find 90% of 72.

7. 25 is what percent of 40?

8. What percent of 50 is 40?

9. Find 35% of 70.

10. 12 is what percent of 20?

SOLUTIONS:

1. $P = R \cdot B$
    $= 65\% \cdot 50$
    $= 0.65 \cdot 50$
    $= 32.5$

2. $R = \dfrac{P}{B}$
    $= \dfrac{25}{40}$
    $= 0.625 = 62.5\%$

3. $B = \dfrac{18}{30\%}$
    $= \dfrac{18}{0.30}$
    $= 60$

4. $R = \dfrac{P}{R}$

$\quad = \dfrac{15}{25}$

$\quad = 0.6 = 60\%$

5. $B = \dfrac{15}{75\%}$

$\quad = \dfrac{15}{0.75}$

$\quad = 20$

6. $P = R \cdot B$

$\quad = 90\% \cdot 72$

$\quad = 0.90 \times 72$

$\quad = 64.8$

7. $R = \dfrac{25}{40}$

$\quad = 0.625 = 62.5\%$

8. $R = \dfrac{P}{B}$

$\quad = \dfrac{40}{50}$

$\quad = 0.80 = 80\%$

9. $P = R \cdot B$

$\quad = 35\% \cdot 70$

$\quad = 0.35 \cdot 70$

$\quad = 24.5$

10. $R = \dfrac{P}{B}$

$\quad = \dfrac{12}{20}$

$\quad = 0.6 = 60\%$

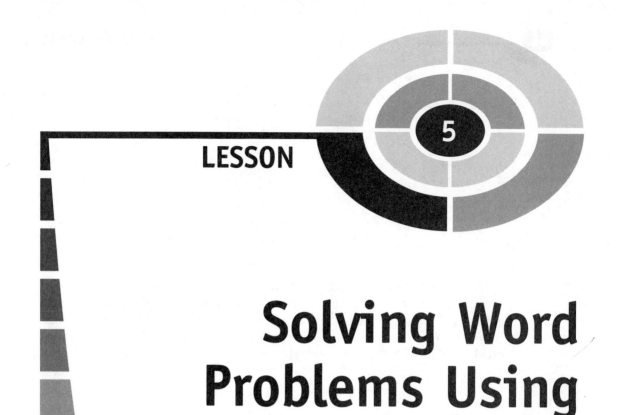

LESSON

5

Solving Word Problems Using Percents

If you need to review percents, complete Refresher III.

A percent problem consists of three values. The base (*B*) is the whole or total, the rate (*R*) is a percent, and the part (*P*). One of these three will be unknown. For example, if a box contains 12 calculators, then the whole is 12. If three calculators are placed on a store's shelf, then 3 is the part. Finally, the percent is $\frac{3}{12} = \frac{1}{4} = 25\%$. That is, 25% of the calculators were placed on the store's shelf.

Percent problems can be solved using the circle method. Figure 5-1 shows how to use the circle method to solve percent problems.

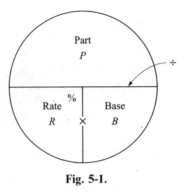

Fig. 5-1.

In the top of the circle, place the part (P). In the lower left portion of the circle, place the rate (%), and in the lower right portion, place the base (B). Now if you are given the bottom two numbers, multiply. That is, $P = R \times B$. If you are given the top number (the part) and one of the bottom numbers, divide to find the other number. That is, $R = \frac{P}{B}$ or $B = \frac{P}{R}$. See Figure 5-2.

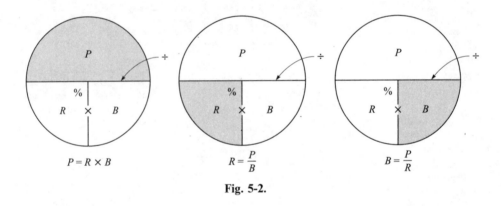

Fig. 5-2.

There are three types of percent problems. They are

Type 1: Finding the part.
Type 2: Finding the rate.
Type 3: Finding the base.

In order to solve percent problems, read the problem and identify the base, rate, and part. One of the three will be the unknown. Substitute the two known quantities in the circle and use the correct formula to find the unknown value. Be sure to change the percent to a decimal before multiplying or dividing.

TYPE 1: FINDING THE PART

In type 1 problems, you are given the base and rate and you are asked to find the part.

EXAMPLE: There are 30 students in a classroom. If 20% of them are absent today, how many students are absent?

SOLUTION:

GOAL: You are being asked to find the number of students who are absent.

STRATEGY: Draw the circle and place 20% in the lower left portion of the circle and 30 in the lower right portion of the circle since it is the total number of students in the class. To find the part, use $P = R \times B$. See Figure 5-3.

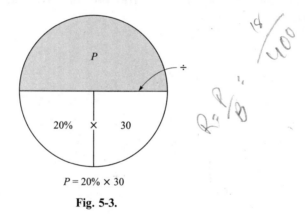

$P = 20\% \times 30$

Fig. 5-3.

IMPLEMENTATION: Substitute in the formula and solve for P.

$P = R \times B$

$P = 20\% \times 30$

$P = 0.20 \times 30$

$P = 6$ students

Hence, 6 students are absent.

EVALUATION: Since $20\% = \frac{1}{5}$ and $\frac{1}{5}$ of $30 = 6$, the answer is correct.

TYPE II: FINDING THE RATE (%)

In type II problems, you are given the part and the whole and are asked to find the rate as a percent. The answer obtained from the formula will be in decimal form. Make sure that you change it into a percent.

EXAMPLE: A person bought a DVD player for $78 and paid a sales tax of $4.68. Find the tax rate.

SOLUTION:

GOAL: You are being asked to find the rate (%).

STRATEGY: In this case, the base (B) is the total cost, which is $78, and the sales tax, $4.68, is the part. Draw the circle and put $78 in the lower right portion and $4.68 in the top portion. To find the rate, use $R = \frac{P}{B}$. See Figure 5-4.

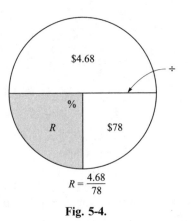

$$R = \frac{4.68}{78}$$

Fig. 5-4.

IMPLEMENTATION:

$$R = \frac{P}{B}$$

$$R = \frac{4.68}{78}$$

$$R = 0.06 = 6\%$$

The sales tax rate is 6%.

EVALUATION: To check the answer, find 6% of $78: 6% × 78 = 0.06 × 78 = $4.68. The answer is correct.

TYPE III: FINDING THE BASE

In type III problems, you are given the part and rate and are asked to find the base or whole.

EXAMPLE: A salesperson earns a 15% commission on all sales. If the commission was $1125, find the amount of his sales.

SOLUTION:

GOAL: You are being asked to find the total amount of sales.

STRATEGY: In this type of problem, you are given the part (commission) and the rate. Place $1125 in the top portion of the circle and the 15% in the bottom left portion. Use $B = \frac{P}{R}$. See Figure 5-5.

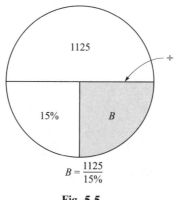

$$B = \frac{1125}{15\%}$$

Fig. 5-5.

IMPLEMENTATION:

$$B = \frac{P}{R}$$

$$B = \frac{1125}{15\%} = \frac{1125}{0.15} = \$7500$$

The total sales were $7500.

EVALUATION: To check the answer, find 15% of $7500: $0.15 \times 7500 = \$1125$. Hence, the answer is correct.

Some percent problems involve finding a percent increase or decrease. Always remember that the original value is used as the base and the amount of the increase or decrease is used as the part. For example, suppose a calculator sold for $40 last week and is on sale for $30 this week. The decrease is $40 − $30 = $10. The percent of decrease is $\frac{10}{40} = 0.25$ or 25%.

EXAMPLE: The price of a textbook increased from $60 to $75. Find the percent increase in the price.

SOLUTION:

GOAL: You are being asked to find the percent of the increase in price.

STRATEGY: Find the increase, and then place that number in the top portion of the circle. The base is the original cost. Use $R = \frac{P}{B}$. See Figure 5-6.

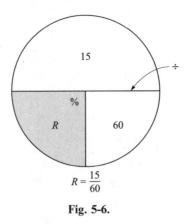

$$R = \frac{15}{60}$$

Fig. 5-6.

IMPLEMENTATION: The increase is $75 − $60 = $15.

$$R = \frac{P}{B}$$

$$R = \frac{15}{60} = \frac{15}{60} = 0.25 = 25\%$$

Hence, the rate of increase is 25%.

EVALUATION: Find 25% of 60: $0.25 \times \$60 = \$15$. The solution is correct.

Try These

1. In a local hospital, there are 80 nurses. If 15% of them are males, how many of the nurses are males?

2. Sean weighed 175 pounds before he lost 28 pounds. Find the percent of his weight loss.

3. The City Garage inspected 110 automobiles. If 90% of them passed, find the number of automobiles that passed.

4. A person paid a 6% sales tax of $29.70 on the purchase of an oven. Find the cost of the oven.

5. In a psychology class, 18% of the students are math majors. If there are 50 students in the class, how many of the students are non-math majors?

6. On an 80-point exam, a student had a score of 70%. How many questions did the student miss?

7. A person bought a classic automobile for $9600 and sold it later for $12,000. What was the percent profit?

8. A video game cost $32 and was discounted $4. What was the discount rate?

9. In a high school, there were 18 male instructors and 32 female instructors. What percent of the instructors are female?

10. An alcoholic beverage comes in 32 ounce bottles and is labeled 10.5% alcohol. How much of the drink is alcohol?

SOLUTIONS:

1. $P = R \times B$

    $P = 15\% \times 80 = 0.15 \times 80 = 12$

2. $R = \dfrac{P}{B}$

    $R = \dfrac{28}{175} = 0.16 = 16\%$

3. $P = R \times B$

   $P = 90\% \times 110 = 0.90 \times 110 = 99$

4. $B = \dfrac{P}{R}$

   $B = \dfrac{\$29.70}{6\%} = \dfrac{\$29.70}{0.06} = \$495$

5. $P = R \times B$

   $P = 18\% \times 50 = 0.18 \times 50 = 9;\ 50 - 9 = 41$

6. $P = R \times B$

   $P = 70\% \times 80 = 0.70 \times 80 = 56;\ 80 - 56 = 24$

7. $\$12{,}000 - \$9600 = \$2400$

   $R = \dfrac{P}{B}$

   $R = \dfrac{\$2400}{\$9600} = 0.25 = 25\%$

8. $R = \dfrac{P}{B}$

   $R = \dfrac{\$4}{\$32} = 0.125 = 12.5\%$

9. $18 + 32 = 50$

   $R = \dfrac{P}{B}$

   $R = \dfrac{32}{50} = 0.64 = 64\%$

10. $P = R \times B$

    $P = 10.5\% \times 32 = 0.105 \times 32 = 3.36$ ounces

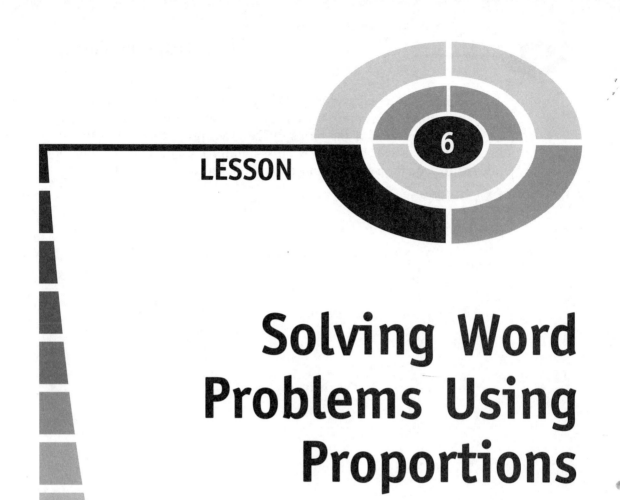

Solving Word Problems Using Proportions

In order to solve word problems using proportions, it is necessary to understand the concept of a ratio. A **ratio** is a comparison of two numbers by using division. For example, the ratio of 4 to 10 is $\frac{4}{10}$, which reduces to $\frac{2}{5}$.

Ratios are used to make comparisons between quantities. If you drive 160 miles in 4 hours, then the ratio of miles to hours is $\frac{160}{4}$ or $\frac{40}{1}$. In other words, you averaged 40 miles per hour.

It is important to understand that whatever number comes first in a ratio statement is placed in the numerator of the fraction and whatever number comes second in the ratio statement is placed in the denominator of the fraction. In general, the ratio of a to b is written as $\frac{a}{b}$.

A **proportion** is a statement of equality of two ratios. For example, $\frac{3}{5} = \frac{9}{15}$ is a proportion. A proportion consists of four terms, and it is usually necessary to find one of the terms of the given proportion given the other three terms. This can be done by cross-multiplying and then dividing both sides of the equation by the numerical coefficient of the variable.

EXAMPLE: Find the value of $x: \dfrac{5}{6} = \dfrac{x}{42}$.

SOLUTION:

$$\frac{5}{6} = \frac{x}{42}$$

$$6 \cdot x = 5 \cdot 42 \qquad \text{Cross-multiply}$$

$$6x = 210$$

$$\frac{\cancel{6}^{1} x}{\cancel{6}^{1}} = \frac{210}{6} \qquad \text{Divide by 6}$$

$$x = 35$$

EXAMPLE: Find the value of $x: \dfrac{10}{x} = \dfrac{30}{6}$.

SOLUTION:

$$\frac{10}{x} = \frac{30}{6}$$

$$30x = 10 \cdot 6 \qquad \text{Cross-multiply}$$

$$30x = 60$$

$$\frac{\cancel{30}^{1} x}{\cancel{30}^{1}} = \frac{60}{30} \qquad \text{Divide by 30}$$

$$x = 2$$

The strategy used to solve problems involving proportions is to identify and write the ratio statement and then write the proportion.

EXAMPLE: If an airplane can travel 360 miles in 3 hours, how far can it travel in 5 hours?

SOLUTION:

GOAL: You are being asked to determine how far the plane can travel in 5 hours.

STRATEGY: Write the ratio: The ratio statement is 360 miles to 3 hours or

$$\frac{360\,\text{miles}}{3\,\text{hours}} = \frac{x\,\text{miles}}{5\,\text{hours}}$$

IMPLEMENTATION: Solve the proportion.

$$\frac{360\,\text{miles}}{3\,\text{hours}} = \frac{x\,\text{miles}}{5\,\text{hours}}$$

$$3x = 360 \cdot 5$$

$$3x = 1800$$

$$\frac{\cancel{3}^1 x}{\cancel{3}^1} = \frac{1800}{3}$$

$$x = 600\,\text{miles}$$

The plane can fly 600 miles.

EVALUATION:

$$\frac{360}{3} = \frac{600}{5}$$ Substitute the value for x in the proportion

$$3 \cdot 600 = 5 \cdot 360$$ and cross-multiply. If the cross products

$$1800 = 1800$$ are equal, the answer is correct.

Notice that when you set up a proportion, always place the same units in the numerator and the same units in the denominator. In the previous problem, $\frac{\text{miles}}{\text{hours}} = \frac{\text{miles}}{\text{hours}}$.

EXAMPLE: If a carpet sells for $15.50 per square yard, how much will it cost to cover a room that measures 20 square yards?

SOLUTION:

GOAL: You are being asked to find the cost of 20 square yards of carpet.

STRATEGY: Write the ratio. It is $\frac{\$15.50}{1 \text{ sq.yd.}}$. The proportion is $\frac{\$15.50}{1 \text{ sq.yd.}} = \frac{x}{20 \text{ sq.yd.}}$.

IMPLEMENTATION: $1 \cdot x = \$15.50 \cdot 20$; $x = \$310$. The cost of the carpet is $310.

EVALUATION:

$$\frac{\$15.50}{1} = \frac{\$310}{20}$$

$$1 \cdot \$310 = \$15.50 \cdot 20$$

$$\$310 = \$310$$

EXAMPLE: If 4 gallons of paint can cover 1120 square feet, how many square feet will 7 gallons of paint cover?

SOLUTION:

GOAL: You are being asked to find how many square feet 7 gallons of paint will cover.

STRATEGY: Write the ratio. It is $\frac{4 \text{ gallons}}{1120 \text{ sq.ft.}}$. The proportion is $\frac{4 \text{ gallons}}{1120 \text{ sq.ft.}} = \frac{7 \text{ gallons}}{x \text{ sq.ft.}}$.

IMPLEMENTATION: Solve the proportion.

$$\frac{4 \text{ gallons}}{1120 \text{ sq.ft.}} = \frac{7 \text{ gallons}}{x \text{ sq.ft.}}$$

$$1120 \cdot 7 = 4 \cdot x$$

$$7840 = 4x$$

$$\frac{7840}{4} = \frac{\cancel{4}^1 x}{\cancel{4}^1}$$

$$1960 \text{ square feet } = x$$

7 gallons will cover 1960 square feet.

EVALUATION:

$$\frac{4}{1120} = \frac{7}{1960}$$

$$1120 \cdot 7 = 4 \cdot 1960$$

$$7840 = 7840$$

EXAMPLE: If the scale on a map reads 2 inches = 50 miles, how many miles are there between two cities whose distance on a map is 7.5 inches?

SOLUTION:

GOAL: You are being asked to find the distance between two cities.

STRATEGY: The ratio statement is $\frac{2 \text{ inches}}{50 \text{ miles}}$. The proportion is $\frac{2 \text{ in.}}{50 \text{ miles}} = \frac{7.5 \text{ in.}}{x \text{ miles}}$.

IMPLEMENTATION:

$$\frac{2}{50} = \frac{7.5}{x}$$

$$50 \cdot 7.5 = 2x$$

$$375 = 2x$$

$$\frac{375}{2} = \frac{\cancel{2}^{1} x}{\cancel{2}^{1}}$$

$$187.5 \text{ miles } = x$$

The cities are 187.5 miles apart.

EVALUATION:

$$\frac{2}{50} = \frac{7.5}{x}$$

$$7.5 \cdot 50 = 2 \cdot 187.5$$

$$375 = 375$$

Try These

1. If a person can burn 120 calories in 15 minutes of cycling, how many calories can the person burn in 75 minutes?

2. If a pizza delivery person drives 276 miles in 3 days, how many miles will the person drive in 5 days?

3. If an author writes 3 chapters in 10 days, how long will it take the author to write a 15-chapter book?

4. If it takes 12 yards to make 3 costumes, how many yards of material will be needed to make 10 costumes?

5. If 4 pounds of grass seed will cover 1250 square feet, how many pounds of grass seed will be needed to cover 3000 square feet?

6. If a person earns $3250 in 3 months, how much will the person earn in 12 months?

7. On a map, the scale is 3 inches = 240 miles. How far apart are two cities whose distance on a map is 8 inches?

8. If two gallons of paint will cover 420 square feet of surface, how many gallons of paint will be needed to cover 1050 square feet?

9. If a recipe calls for 3.5 cups of flour to serve 4 people, how many cups of flour will be needed to make the recipe serve 9 people?

10. If a merchant can order 5 pairs of the same kind of shoes for $89, how much will 12 pairs cost?

SOLUTIONS:

1. $$\frac{120 \text{ calories}}{15 \text{ min}} = \frac{x \text{ calories}}{75 \text{ min}}$$

$$15x = 120 \cdot 75$$

$$15x = 9000$$

$$\frac{\cancel{15}^{1} x}{\cancel{15}^{1}} = \frac{9000}{15}$$

$$x = 600 \text{ calories}$$

2. $$\frac{276 \text{ miles}}{3 \text{ days}} = \frac{x \text{ miles}}{5 \text{ days}}$$

$$3x = 276 \cdot 5$$

$$3x = 1380$$

$$\frac{\cancel{3}^{1} x}{\cancel{3}^{1}} = \frac{1380}{3}$$

$$x = 460 \text{ miles}$$

3. $\dfrac{3 \text{ chapters}}{10 \text{ days}} = \dfrac{15 \text{ chapters}}{x \text{ days}}$

$$10 \cdot 15 = 3x$$

$$150 = 3x$$

$$\dfrac{150}{3} = \dfrac{\cancel{3}^1 x}{\cancel{3}^1}$$

$$50 \text{ days} = x$$

4. $\dfrac{12 \text{ yards}}{3 \text{ costumes}} = \dfrac{x \text{ yards}}{10 \text{ costumes}}$

$$3x = 12 \cdot 10$$

$$3x = 120$$

$$\dfrac{\cancel{3}^1 x}{\cancel{3}^1} = \dfrac{120}{3}$$

$$x = 40 \text{ yards}$$

5. $\dfrac{4 \text{ pounds}}{1250 \text{ sq.ft.}} = \dfrac{x \text{ pounds}}{3000 \text{ sq.ft.}}$

$$1250x = 4 \cdot 3000$$

$$1250x = 12,000$$

$$\dfrac{\cancel{1250}^1 x}{\cancel{1250}^1} = \dfrac{12,000}{1250}$$

$$x = 9.6 \text{ pounds}$$

6. $\dfrac{\$3250}{3 \text{ months}} = \dfrac{x}{12 \text{ months}}$

$$3x = 3250 \cdot 12$$

$$3x = 39,000$$

$$\dfrac{\cancel{3}^1 x}{\cancel{3}^1} = \dfrac{39,000}{3}$$

$$x = \$13,000$$

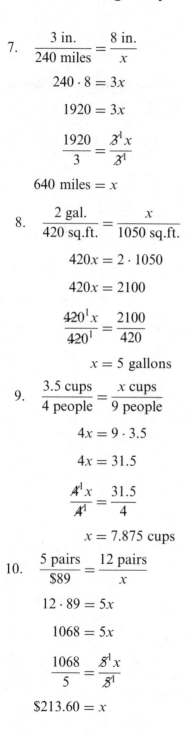

7. $\dfrac{3 \text{ in.}}{240 \text{ miles}} = \dfrac{8 \text{ in.}}{x}$

$240 \cdot 8 = 3x$

$1920 = 3x$

$\dfrac{1920}{3} = \dfrac{3^1 x}{3^1}$

$640 \text{ miles} = x$

8. $\dfrac{2 \text{ gal.}}{420 \text{ sq.ft.}} = \dfrac{x}{1050 \text{ sq.ft.}}$

$420x = 2 \cdot 1050$

$420x = 2100$

$\dfrac{420^1 x}{420^1} = \dfrac{2100}{420}$

$x = 5 \text{ gallons}$

9. $\dfrac{3.5 \text{ cups}}{4 \text{ people}} = \dfrac{x \text{ cups}}{9 \text{ people}}$

$4x = 9 \cdot 3.5$

$4x = 31.5$

$\dfrac{4^1 x}{4^1} = \dfrac{31.5}{4}$

$x = 7.875 \text{ cups}$

10. $\dfrac{5 \text{ pairs}}{\$89} = \dfrac{12 \text{ pairs}}{x}$

$12 \cdot 89 = 5x$

$1068 = 5x$

$\dfrac{1068}{5} = \dfrac{5^1 x}{5^1}$

$\$213.60 = x$

LESSON 7

Solving Word Problems Using Formulas

In mathematics and science, many problems can be solved by using a formula. A **formula** is a mathematical statement of the relationship of two or more variables. For example, the distance (D) an automobile travels is related to the rate (R) of speed and the time (T) it travels. In symbols, $D = RT$. In order to evaluate formulas, you use the order of operations.

Step 1: Perform all operations in parentheses.
Step 2: Raise each number to its power.
Step 3: Perform multiplication and division from left to right.
Step 4: Perform addition and subtraction from left to right.

EXAMPLE: Find the interest on a loan whose principal (P) is \$4800 at a rate of 5% for 6 years. Use $I = PRT$.

SOLUTION:

GOAL: You are being asked to find the interest.

STRATEGY: Use the formula $I = PRT$.

IMPLEMENTATION:

$$I = PRT$$

$$= \$4800 \cdot 5\% \cdot 6$$

$$= \$4800 \cdot (0.05) \cdot 6$$

$$= \$1440$$

EVALUATION: You can estimate the answer by rounding \$4800 to \$5000 and then finding 5% of \$5000, which is $0.05 \times \$5000 = \$250$. The interest for one year is about \$250. The interest for 6 years then is $6 \times \$250 = \$1500$. Since this is close to \$1440, the answer is probably correct.

EXAMPLE: Find the Celsius temperature (C) when the Fahrenheit temperature (F) is 86°. Use $C = \frac{5}{9}(F - 32°)$.

SOLUTION:

GOAL: You are being asked to find a Celsius temperature.

STRATEGY: Use the formula $C = \frac{5}{9}(F - 32°)$.

IMPLEMENTATION:

$$C = \frac{5}{9}(F - 32°).$$

$$C = \frac{5}{9}(86 - 32)$$

$$= \frac{5}{9}(54)$$

$$= 30°$$

EVALUATION: You can estimate the answer by subtracting 30° from the Fahrenheit temperature and dividing the answer by 2: $86 - 30 = 56$; $56 \div 2 = 28°$. Since 28° is close to 30°, the answer is probably correct.

EXAMPLE: The distance (d) an object falls in feet is $d = 32t^2$ where t is the time in seconds. Find the distance an object falls in 3 seconds.

SOLUTION:

GOAL: You are being asked to find the distance an object falls in 3 seconds.

STRATEGY: Use the formula $d = 32t^2$.

IMPLEMENTATION:

$$d = 32t^2$$
$$= 32(3)^2$$
$$= 32 \cdot 9$$
$$= 288 \text{ feet}$$

EVALUATION: Estimate the answer by rounding 32 to 30 and then multiplying $30 \times 9 = 270$. Since this estimate is close to 288, the answer is probably correct.

Try These

1. Find the amount of work (W) done by applying a force (F) of 60 pounds moving a distance (d) of 9 feet. Use $W = Fd$.

2. Find the distance (D) an automobile travels at a rate (R) of 35 miles per hour in 2.5 hours (T). Use $D = RT$.

3. Find the amount of interest (I) earned on a principal (P) of $6000 at a rate ($R$) of 7% for a time ($T$) of 8 years. Use $I = PRT$.

4. Find the perimeter (P) of a rectangle whose length (l) is 20 inches and whose width (w) is 8 inches. Use $P = 2l + 2w$.

5. Find the Celsius temperature (C) when the Fahrenheit temperature (F) is 50°. Use $C = \frac{5}{9}(F - 32°)$.

6. Find the volume (V) of a cylinder in cubic feet when the height (h) is 19 feet and the radius (r) of the base is 3 feet. Use $V = 3.14r^2h$.

7. Find the force (F) of the wind against a flat surface whose area (A) is 20 square feet when the wind speed (s) is 35 miles per hour. Use $F = 0.004As^2$.

8. Find the surface area (A) of a cube in square feet when each side (s) measures 12 inches. Use $A = 6s^2$.

9. Find the current (I) in amperes when the electromotive force (E) is 18 volts and the resistance (R) is 6 ohms. Use $I = \frac{E}{R}$.

10. Find the Fahrenheit temperature (F) when the Celsius temperature (C) is 50°. Use $F = \frac{9}{5}C + 32°$.

SOLUTIONS:

1. $W = Fd$

   $= 60 \cdot 9$

   $= 540$ ft.-lb.

2. $D = RT$

   $= 35 \cdot 2.5$

   $= 87.5$ miles

3. $I = PRT$

   $= \$6000 \cdot 7\% \cdot 8$

   $= \$6000 \cdot (0.07) \cdot 8$

   $= \$3360$

4. $P = 2l + 2w$

$= 2 \cdot 20 + 2 \cdot 8$

$= 40 + 16$

$= 56$ in.

5. $C = \dfrac{5}{9}(F - 32°)$

$= \dfrac{5}{9}(50 - 32°)$

$= \dfrac{5}{9} \cdot 18$

$= 10°$

6. $V = 3.14r^2h$

$= 3.14 \cdot 3^2 \cdot 19$

$= 3.14 \cdot 9 \cdot 19$

$= 536.94$ cubic feet

7. $F = 0.004As^2$

$= 0.004 \cdot 20 \cdot 35^2$

$= 0.004 \cdot 20 \cdot 1225$

$= 98$ pounds

8. $A = 6s^2$

$= 6 \cdot 12^2$

$= 6 \cdot 144$

$= 864$ square inches

9. $I = \dfrac{E}{R}$

    $= \dfrac{18}{6}$

    $= 3$ amperes

10. $F = \dfrac{9}{5}C + 32°$

    $= \dfrac{9}{5} \cdot 50 + 32$

    $= 90 + 32$

    $= 122°$

Quiz 2

1. Find the sales tax on a book that costs $25 if the rate is 6%.
   - (a) $15
   - (b) $0.15
   - (c) $150
   - (d) $1.50 ╱

2. If a person earned $32,000 a year and received an $800 raise, what was the percent increase in her salary?
   - (a) 25%
   - (b) 2.5%
   - (c) 0.25%
   - (d) 250%

3. A sales person received a commission of $60 on a sale of an item. If the commission rate is 20%, find the amount of the sale.
   - (a) $300 ╱
   - (b) $12
   - (c) $120
   - (d) $30

4. If a DVD player originally sold for $80 and was reduced 30% for a sale, what was the reduced price?
   (a) $24
   (b) $240
   (c) $56
   (d) $5.60

5. If a family purchased a home for $120,000 and put 20% down, how much was left to finance?
   (a) $24,000
   (b) $42,000
   (c) $69,000
   (d) $96,000

6. If a person can type 10 pages of text in 4 minutes, how many pages can the person type in 25 minutes?
   (a) 40
   (b) 25
   (c) 62.5
   (d) 100

$$\frac{10P}{4m} = \frac{xP}{25}$$

7. If three electrical switches cost $2.10, how much will 10 switches cost?
   (a) $7.00
   (b) $21.00
   (c) $10.00
   (d) $6.30

8. If 2 ounces of a food contains 130 calories, how many calories would be contained in 7 ounces?
   (a) 260
   (b) 455
   (c) 910
   (d) 1400

$$\frac{2}{130} = \frac{7}{x}$$

9. If a person travels a distance of 840 miles in 12 hours, find the rate. Use $R = \frac{D}{T}$.
   - (a) 50 miles per hour
   - (b) 40 miles per hour
   - (c) 60 miles per hour
   - (d) 70 miles per hour ✓

10. How far (in feet) will an object fall in 5 seconds? Use $d = 32t^2$.
    - (a) 160 ft.
    - (b) 25 ft. ✓
    - (c) 800 ft.
    - (d) 650 ft.

Equations

An **algebraic expression** consists of variables (letters), numbers, operation signs ($+$, $-$, $\times$, $\div$), and grouping symbols. Here are a few examples of algebraic expressions:

$$3x \qquad 5(x-5) \qquad -8x^2 \qquad 9+2$$

An **equation** is a statement of equality of two algebraic expressions. Here are some examples of equations:

$$5+4=9 \qquad 3x-2=13 \qquad x^2+3x+2=0$$

An equation which contains a variable is called a **conditional** equation. To **solve** a conditional equation, it is necessary to find a number which, when substituted for the variable, makes a true equation. This number is called a **solution** to the equation. For example, 5 is a solution to the equation $x+3=8$, since when 5 is substituted for x it makes the equation true, i.e., $5+3=8$. The process of finding a solution to an equation is called **solving** the equation.

The same number (except zero) can be added to, subtracted from, multiplied by, or divided into both sides of the equation without changing the nature of the equation.

EXAMPLE: Solve for x: $x - 7 = 9$

SOLUTION:

$$x - 7 = 9$$

$$x - 7 + 7 = 9 + 7 \qquad \text{add 7 to both sides}$$

$$x = 16$$

EXAMPLE: Solve for x: $x + 8 = 10$

SOLUTION:

$$x + 8 = 10$$

$$x + 8 - 8 = 10 - 8 \qquad \text{subtract 8 from both sides}$$

$$x = 2$$

EXAMPLE: Solve: $\dfrac{x}{4} = 5$

SOLUTION:

$$\frac{x}{4} = 5$$

$$\frac{\cancel{4}^1}{1} \cdot \frac{x}{\cancel{4}^1} = 4 \cdot 5 \qquad \text{multiply both sides by 4}$$

$$x = 20$$

EXAMPLE: Solve: $6x = 42$

SOLUTION:

$$6x = 42$$

$$\frac{\cancel{6}^1 x}{\cancel{6}^1} = \frac{42}{6}$$

$$x = 7$$

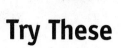
Try These

Solve each equation for x:

1. $x + 18 = 40$
2. $x - 22 = 10$
3. $5x = 45$
4. $x + 6 = 16$
5. $\dfrac{x}{3} = 8$

SOLUTIONS:

1. $$x + 18 = 40$$
$$x + 18 - 18 = 40 - 18$$
$$x = 22$$

2. $$x - 22 = 10$$
$$x - 22 + 22 = 10 + 22$$
$$x = 32$$

3. $$5x = 45$$
$$\frac{\cancel{5}^{1}x}{\cancel{5}^{1}} = \frac{45}{5}$$
$$x = 9$$

4. $$x + 6 = 16$$
$$x + 6 - 6 = 16 - 6$$
$$x = 10$$

5. $$\frac{x}{3} = 8$$
$$\frac{\cancel{3}^{1}}{1} \cdot \frac{x}{\cancel{3}^{1}} = 8 \cdot 3$$
$$x = 24$$

More complex equations require several steps to solve them. The goal is to use addition and/or subtraction to get the variables on one side of the equation and the numbers on the other side of the equation. Then divide both sides by the number in front of the variables.

EXAMPLE: Solve for x: $4x + 10 = 34$

SOLUTION:

$$4x + 10 = 34$$

$$4x + 10 - 10 = 34 - 10 \qquad \text{subtract 10}$$

$$4x = 24$$

$$\frac{\cancel{4}^1 x}{\cancel{4}^1} = \frac{24}{4} \qquad \text{divide by 4}$$

$$x = 6$$

EXAMPLE: Solve for x: $5x - 6 = 3x + 14$

SOLUTION:

$$5x - 6 = 3x + 14$$

$$5x - 6 + 6 = 3x + 14 + 6 \qquad \text{add 6}$$

$$5x = 3x + 20$$

$$5x - 3x = 3x - 3x + 20 \qquad \text{subtract } 3x$$

$$2x = 20$$

$$\frac{\cancel{2}^1 x}{\cancel{2}^1} = \frac{20}{2} \qquad \text{divide by 2}$$

$$x = 10$$

An equation can be **checked** by substituting the solution back into the original equation and seeing if a true equation results. To check the equation in the previous example, substitute 10 for x in the original equation.

$$5x - 6 = 3x + 14$$
$$5(10) - 6 = 3(10) + 14$$
$$50 - 6 = 30 + 14$$
$$44 = 44$$

Try These

Solve each equation for x:

1. $7x + 3 = 31$
2. $2x - 14 = 16$
3. $12x - 10 = 8x + 26$
4. $6x + 5 = 3x + 32$
5. $10x - 7 = 5x + 33$

SOLUTIONS:

1. $$7x + 3 = 31$$
$$7x + 3 - 3 = 31 - 3$$
$$7x = 28$$
$$\frac{\cancel{7}^1 x}{\cancel{7}^1} = \frac{28}{7}$$
$$x = 4$$

2. $$2x - 14 = 16$$
$$2x - 14 + 14 = 16 + 14$$
$$2x = 30$$
$$\frac{\cancel{2}^1 x}{\cancel{2}^1} = \frac{30}{2}$$
$$x = 15$$

3.
$$12x - 10 = 8x + 26$$
$$12x - 10 + 10 = 8x + 26 + 10$$
$$12x - 8x = 8x - 8x + 36$$
$$4x = 36$$
$$\frac{\cancel{4}^1 x}{\cancel{4}^1} = \frac{36}{4}$$
$$x = 9$$

4.
$$6x + 5 = 3x + 32$$
$$6x + 5 - 5 = 3x + 32 - 5$$
$$6x = 3x + 27$$
$$6x - 3x = 3x - 3x + 27$$
$$3x = 27$$
$$\frac{\cancel{3}^1 x}{\cancel{3}^1} = \frac{27}{3}$$
$$x = 9$$

5.
$$10x - 7 = 5x + 33$$
$$10x - 7 + 7 = 5x + 33 + 7$$
$$10x = 5x + 40$$
$$10x - 5x = 5x - 5x + 40$$
$$5x = 40$$
$$\frac{\cancel{5}^1 x}{\cancel{5}^1} = \frac{40}{5}$$
$$x = 8$$

Many equations contain parentheses. In order to remove parentheses, multiply each term inside the parentheses by the number outside the parentheses. For example,

$$3(x + 2) = 3 \cdot x + 3 \cdot 2 = 3x + 6$$
$$5(2x - 8) = 5 \cdot 2x - 5 \cdot 8 = 10x - 40$$

When you solve an equation, remove parentheses first, combine like terms (i.e., $6x + 8x$), and then solve as shown in the previous examples.

EXAMPLE: Solve for x: $2(3x - 12) = 36$

SOLUTION:

$$2(3x - 12) = 36$$
$$2 \cdot 3x - 2 \cdot 12 = 36 \qquad \text{remove parentheses}$$
$$6x - 24 = 36$$
$$6x - 24 + 24 = 36 + 24 \quad \text{add 24}$$
$$6x = 60$$
$$\frac{\cancel{6}^1 x}{\cancel{6}^1} = \frac{60}{6} \qquad \text{divide by 6}$$
$$x = 10$$

EXAMPLE: Solve for x: $5(3x + 8) - 10x = 35$

SOLUTION:

$$5(3x + 8) - 10x = 35$$
$$15x + 40 - 10x = 35 \qquad \text{remove parentheses}$$
$$5x + 40 = 35 \qquad \text{combine } 15x - 10x$$
$$5x + 40 - 40 = 35 - 40 \quad \text{subtract 40}$$
$$5x = -5$$
$$\frac{\cancel{5}^1 x}{\cancel{5}^1} = -\frac{5}{5} \qquad \text{divide by 5}$$
$$x = -1$$

Try These

1. $3(x - 6) + 10 = 16$

2. $7(2x - 5) = 7x + 7$

3. $5(3x - 10) = 25$

4. $3(x - 8) = 2(x + 5)$

5. $4(3x + 2) = 10x - 14$

SOLUTIONS:

1.
$$3(x - 6) + 10 = 16$$
$$3x - 18 + 10 = 16$$
$$3x - 8 = 16$$
$$3x - 8 + 8 = 16 + 8$$
$$3x = 24$$
$$\frac{\cancel{3}^1 x}{\cancel{3}^1} = \frac{24}{3}$$
$$x = 8$$

2.
$$7(2x - 5) = 7x + 7$$
$$14x - 35 = 7x + 7$$
$$14x - 7x - 35 = 7x - 7x + 7$$
$$7x - 35 = 7$$
$$7x - 35 + 35 = 7 + 35$$
$$7x = 42$$
$$\frac{\cancel{7}^1 x}{\cancel{7}^1} = \frac{42}{7}$$
$$x = 6$$

3.

$$5(3x - 10) = 25$$

$$15x - 50 = 25$$

$$15x - 50 + 50 = 25 + 50$$

$$15x = 75$$

$$\frac{\cancel{15}^1 x}{\cancel{15}^1} = \frac{75}{15}$$

$$x = 5$$

4.

$$3(x - 8) = 2(x + 5)$$

$$3x - 24 = 2x + 10$$

$$3x - 2x - 24 = 2x - 2x + 10$$

$$x - 24 = 10$$

$$x - 24 + 24 = 10 + 24$$

$$x = 34$$

5.

$$4(3x + 2) = 10x - 14$$

$$12x + 8 = 10x - 14$$

$$12x - 10x + 8 = 10x - 10x - 14$$

$$2x + 8 = -14$$

$$2x + 8 - 8 = -14 - 8$$

$$2x = -22$$

$$\frac{\cancel{2}^1 x}{\cancel{2}^1} = \frac{-22}{2}$$

$$x = -11$$

Sometimes when you are solving word problems, you will need to solve an equation containing fractions. It should be noted that fraction terms can be written in two ways. See the next examples:

$$\frac{1}{3}x \text{ can be written as } \frac{x}{3}$$

$$\frac{5}{6}x \text{ can be written as } \frac{5x}{6}$$

$$\frac{3}{4}(x-2) \text{ can be written as } \frac{3(x-2)}{4}$$

To solve an equation containing fractions, it is necessary to find the lowest common denominator of all the fractions, and then multiply each term in the equation by the lowest common denominator. This process is called **clearing fractions**.

EXAMPLE: Solve for x: $\dfrac{x}{2}+\dfrac{x}{3}=10$

SOLUTION:

$$\frac{x}{2}+\frac{x}{3}=10 \qquad \text{the LCD is 6}$$

$$\frac{\cancel{6}^3}{1}\cdot\frac{x}{\cancel{2}_1}+\frac{\cancel{6}^2}{1}\cdot\frac{x}{\cancel{3}_1}=6\cdot10 \qquad \text{clear fractions}$$

$$3x+2x=60$$

$$5x=60 \qquad \text{combine terms}$$

$$\frac{\cancel{5}^1 x}{\cancel{5}^1}=\frac{60}{5} \qquad \text{divide by 5}$$

$$x=12$$

EXAMPLE: Solve for x: $\dfrac{1}{6}+\dfrac{1}{3}=\dfrac{1}{x}$

SOLUTION:

$$\frac{1}{6}+\frac{1}{3}=\frac{1}{x}$$ the LCD is $6x$

$$\frac{\cancel{6}^1 x}{1}\cdot\frac{1}{\cancel{6}^1}+\frac{\cancel{6}^2 x}{1}\cdot\frac{1}{\cancel{3}^1}=\frac{6\cancel{x}^1}{1}\cdot\frac{1}{\cancel{x}^1}$$ clear fractions

$$x+2x=6$$ combine terms

$$3x=6$$

$$\frac{\cancel{3}^1 x}{\cancel{3}^1}=\frac{6}{3}$$ divide by 3

$$x=2$$

Try These

Solve each equation for x:

1. $\dfrac{x}{2}+\dfrac{x}{3}+\dfrac{x}{4}=26$

2. $\dfrac{2}{3}x+\dfrac{3}{4}=\dfrac{5}{8}$

3. $\dfrac{1}{5}+\dfrac{1}{6}=\dfrac{1}{x}$

4. $\dfrac{1}{2}(x-3)=\dfrac{2}{3}$

5. $\dfrac{1}{4}x+20=\dfrac{3}{8}x$

SOLUTIONS:

1.
$$\frac{x}{2} + \frac{x}{3} + \frac{x}{4} = 26$$

$$\frac{\cancel{12}^{6}}{1} \cdot \frac{x}{\cancel{2}^{1}} + \frac{\cancel{12}^{4}}{1} \cdot \frac{x}{\cancel{3}^{1}} + \frac{\cancel{12}^{3}}{1} \cdot \frac{x}{\cancel{4}^{1}} = 26 \cdot 12$$

$$6x + 4x + 3x = 312$$

$$\frac{\cancel{13}^{1}x}{\cancel{13}^{1}} = \frac{312}{13}$$

$$x = 24$$

2.
$$\frac{2}{3}x + \frac{3}{4} = \frac{5}{8}$$

$$\frac{\cancel{24}^{8}}{1} \cdot \frac{2}{\cancel{3}^{1}}x + \frac{\cancel{24}^{6}}{1} \cdot \frac{3}{\cancel{4}^{1}} = \frac{\cancel{23}^{3}}{1} \cdot \frac{5}{\cancel{8}^{1}}$$

$$16x + 18 = 15$$

$$16x + 18 - 18 = 15 - 18$$

$$16x = -3$$

$$\frac{\cancel{16}^{1}x}{\cancel{16}^{1}} = -\frac{3}{16}$$

$$x = -\frac{3}{16}$$

3.
$$\frac{1}{5}+\frac{1}{6}=\frac{1}{x}$$

$$\frac{\cancel{30}^6 x}{1}\cdot\frac{1}{\cancel{5}^1}+\frac{\cancel{30}^5 x}{1}\cdot\frac{1}{\cancel{6}^1}=30\cancel{x}^1\cdot\frac{1}{\cancel{x}^1}$$

$$6x+5x=30$$

$$11x=30$$

$$\frac{\cancel{11}^1 x}{\cancel{11}^1}=\frac{30}{11}$$

$$x=\frac{30}{11}=2\frac{8}{11}$$

4.
$$\frac{1}{2}(x-3)=\frac{2}{3}$$

$$\frac{\cancel{6}^3}{1}\cdot\frac{1}{\cancel{2}^1}(x-3)=\frac{\cancel{6}^2}{1}\cdot\frac{2}{\cancel{3}^1}$$

$$3(x-3)=4$$

$$3x-9=4$$

$$3x-9+9=4+9$$

$$3x=13$$

$$\frac{\cancel{3}^1 x}{\cancel{3}^1}=\frac{13}{3}$$

$$x=4\frac{1}{3}$$

5.
$$\frac{1}{4}x+20=\frac{3}{8}x$$

$$\frac{\cancel{8}^2}{1}\cdot\frac{1}{\cancel{4}^1}x+8\cdot20=\frac{\cancel{8}^1}{1}\cdot\frac{3}{\cancel{8}^1}x$$

$$2x+160=3x$$

$$2x-2x+160=3x-2x$$

$$160=x$$

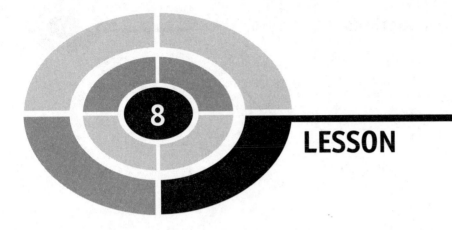

LESSON

Algebraic Representation

When you solve an algebra word problem, you must first be able to translate the conditions of the problem into an equation involving algebraic expressions. An algebraic expression will consist of variables (letters), numbers, operation signs ($+$, $-$, $\times$, $\div$), and grouping symbols such as parentheses.

Here are some common phrases that are used in algebra word problems: **Addition** can be denoted by

> sum
> added to
> increased by
> larger than
> more than

Subtraction can be denoted by

less than
subtracted from
decreased by

s are translated into symbols:

Symbolic Representation

$5x$

$x + 3$

$x - 7$

$\frac{1}{2}x$ or $\frac{x}{2}$

x^2

Nine added to twice a number x $9 + 2x$

Four times a number minus eight $4x - 8$

The cost of x feet of rope at 15 cents a foot $0.15x$

Try These

Write each in symbols:

1. Six times a number x decreased by four.
2. A number x increased by six.
3. Eight less than a number x.
4. Fifteen minus one-third of a number x.
5. The square of a number x plus 7.

SOLUTIONS:

1. $6x - 4$
2. $x + 6$
3. $x - 8$
4. $15 - \frac{1}{3}x$ or $15 - \frac{x}{3}$
5. $x^2 + 7$

In the previous examples, only one unknown was used. At other times, it is necessary to represent two *related* unknowns by using one variable. Consider these examples:

"*The sum of two numbers is 12.*" When you are given two numbers whose sum is 12 and one number is, say, 7, how would you find the other number? You would say $12 - 7$. So if one number is x, the other number would be $12 - x$.

"*One number is six more than another number.*" If I told you one number is 10, how would you find the other number? You would add $10 + 6$. So if one stated number is x, the other number would be $x + 6$.

"*One number is five times another number.*" If I told you one number is two, how would you find the other number? You would multiply two by five. So if one number is x, the other number would be $5x$.

"*One number is twelve less than another number.*" If I told you one number was 30, how would you find the other number? You would subtract $30 - 12$. So if one number is x, the other number is $x - 12$.

Try These

Represent each using symbols:

1. The sum of two numbers is 15.

2. One number is four times another number.
3. One number is 6 more than twice the other number.
4. One number is 4 less than one-half another number.
5. The difference of two numbers is 12.

SOLUTIONS:

1. Let $x =$ the first number and $15 - x =$ the second number.
2. Let $x =$ the first number and $4x =$ the second number.
3. Let $x =$ the first number and $2x + 6 =$ the second number.
4. Let $x =$ the first number and $\frac{1}{2}x - 4 =$ the second number.
5. Let $x =$ the first number and $x - 12 =$ the second number.

The third skill necessary to solve word problems is to be able to translate the given symbols of representation into an equation. Consider these examples.

"Three times a number increased by 10 is equal to 28" translates to

$$3 \cdot x + 10 = 28$$

"Seven times a number decreased by 6 is equal to 29."

$$7 \cdot x - 6 = 29$$

"The difference between a number and one-half itself is equal to 12."

$$x - \frac{1}{2}x = 12$$

Try These

Translate each into an equation:

1. Seven increased by four times a number is 31.
2. If 5 is subtracted from two times a number, the answer is 15.
3. One half a number plus 8 is equal to 19.
4. Two times a number plus $1.20 is equal to $1.80.
5. The sum of a number and two times itself is equal to 30.

SOLUTIONS:

1. $7+4x=31$
2. $2x-5=15$
3. $\frac{1}{2}x+8=19$
4. $2x+1.20=1.80$
5. $x+2x=30$

Finally, it is necessary to be able to write an equation for two related unknowns using one variable.

EXAMPLE: Write an equation for this problem: "One number is 10 more than another number and their sum is 16."

SOLUTION: Let $x=$ the smaller number and $x+10=$ the larger number. The equation is $x+x+10=16$.

EXAMPLE: Write an equation for this problem: "One number is five times as large as another number. If three times the smaller number is subtracted from the larger number, the answer is 20."

SOLUTION: Let $x=$ the smaller number and $5x=$ the larger number. The equation is $5x-3x=20$.

Try These

Write an equation for each:

1. The smaller number is $\frac{1}{2}$ of the larger number. Find the numbers if their sum is 36.

2. A certain number exceeds another number by 6. If their sum is 56, find the numbers.

3. One number is 8 more than twice another number. Find the numbers if their sum is 50.

4. What number increased by $\frac{1}{4}$ of itself is equal to 5?

5. Two times a number is 6 more than $\frac{1}{2}$ the number. Find the numbers.

SOLUTIONS:

1. Let $x =$ the larger number and $\frac{1}{2}x =$ the smaller number. The equation is $x + \frac{1}{2}x = 36$.

2. Let $x =$ the smaller number and $x + 6 =$ the larger number. The equation is $x + x + 6 = 56$.

3. Let $x =$ the smaller number and $2x + 8 =$ the other number. The equation is $x + 2x + 8 = 50$.

4. Let $x =$ the number. The equation is $x + \frac{1}{4}x = 5$.

5. Let $x =$ the number. The equation is $2x = \frac{1}{2}x + 6$.

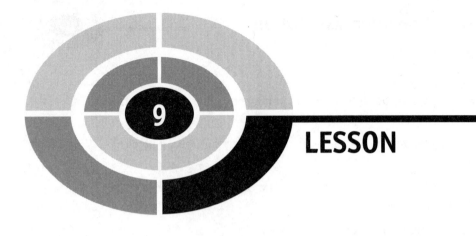

Solving Number Problems

The strategy that is used to solve word problems in algebra is as follows:

1. Represent an unknown by using x.
2. If necessary, represent the other unknowns by using algebraic expressions in terms of x.
3. From the conditions of the problem, write an equation using the algebraic representation of the unknown(s).

EXAMPLE: One number is 8 more than another number and the sum of the two numbers is 26. Find the numbers.

SOLUTION:

GOAL: You are being asked to find two numbers.

STRATEGY: Let $x =$ the smaller number and $x+8 =$ the larger number. Since the problem asked for the sum, write the equation as $x + x + 8 = 26$.

IMPLEMENTATION: Solve the equation for x:

$$x + x + 8 = 26$$

$$2x + 8 = 26$$

$$2x + 8 - 8 = 26 - 8$$

$$2x = 18$$

$$\frac{2^1 x}{2^1} = \frac{18}{2}$$

$$x = 9$$

Hence one number is 9 and the other number is $x+8$ or $9+8=17$.

EVALUATION: Check the answer: $9+17=26$

EXAMPLE: If eight plus three times a number is equal to 23, find the number.

SOLUTION:

GOAL: You are being asked to find one number.

STRATEGY: Let $x =$ the number. Eight plus three times a number is written as $8 + 3x$ and the equation is $8 + 3x = 23$.

IMPLEMENTATION: Solve the equation.

$$8 + 3x = 23$$

$$8 - 8 + 3x = 23 - 8$$

$$3x = 15$$

$$\frac{3^1 x}{3^1} = \frac{15}{3}$$

$$x = 5$$

LESSON 9 Solving Number Problems

EVALUATION: Check the answer: $8 + 3 \cdot 5 = 8 + 15$

$$= 23$$

EXAMPLE: A psychology class with 52 people is divided into two sections so that there are four more students in one section than the other. How many students were in each section?

SOLUTION:

GOAL: You are being asked to find the number of students in two sections.

STRATEGY: Let $x =$ the number of students in one section and $x + 4$ be the number of students in the other section. The equation is $x + x + 4 = 52$.

IMPLEMENTATION: Solve the equation.

$$x + x + 4 = 52$$

$$2x + 4 = 52$$

$$2x + 4 - 4 = 52 - 4$$

$$2x = 48$$

$$\frac{2^1 x}{2^1} = \frac{48}{2}$$

$$x = 24$$

$$x + 4 = 28$$

EVALUATION: Check the answer: $24 + 28 = 52$

EXAMPLE: The sum of two numbers is 16. If five times the first number is equal to three times the second number, find the numbers.

SOLUTION:

GOAL: You are being asked to find two numbers whose sum is 16.

STRATEGY: Let $x =$ the first number and $(16 - x) =$ the second number.

Now five times the first number is $5x$ and 3 times the second number is $3(16 - x)$. The equation is $5x = 3(16 - x)$.

IMPLEMENTATION: Solve the equation.

$$5x = 3(16 - x)$$

$$5x = 48 - 3x$$

$$5x + 3x = 48 - 3x + 3x$$

$$8x = 48$$

$$\frac{8^1 x}{8^1} = \frac{48}{8}$$

$$x = 6$$

$$16 - x = 16 - 6 = 10$$

Hence the first number is 6 and the second number is 10.

EVALUATION: Check the answer: $6 + 10 = 16$ and $5 \cdot 6 = 3 \cdot 10$ or $30 = 30$

Some number problems use **consecutive integers**. Numbers such as 1, 2, 3, 4, 5, etc., are called consecutive integers. They differ by one. Consecutive integers can be represented as:

Let $x =$ the first integer
$x + 1 =$ the second integer
$x + 2 =$ the third integer, etc.

Consecutive odd integers are numbers such as 1, 3, 5, 7, 9, 11, etc. They differ by two. They can be represented as:

Let $x =$ the first odd integer
$x + 2 =$ the second consecutive integer
$x + 4 =$ the third consecutive odd integer, etc.

Consecutive even integers are numbers such as 2, 4, 6, 8, 10, 12, etc. They also differ by two. They can be represented as:

Let $x =$ the first even integer
$x + 2 =$ the second consecutive even integer
$x + 4 =$ the third consecutive even integer, etc.

LESSON 9 Solving Number Problems

You need not worry whether you are looking for consecutive even or odd numbers since the problems will always work out correctly. (Textbook authors have made them up so that they will.)

EXAMPLE: Find three consecutive integers whose sum is 63.

SOLUTION:

GOAL: You are being asked to find three consecutive numbers whose sum is 63.

STRATEGY: Let $x=$ the first integer and $x+1=$ the second integer and $x+2=$ the third integer. The equation is $x+(x+1)+(x+2)=63$

IMPLEMENTATION: Solve the equation:

$$x+x+1+x+2=63$$

$$3x+3=63$$

$$3x+3-3=63-3$$

$$x=20$$

$$x+1=21$$

$$x+2=22$$

EVALUATION: Check the answer: $20+21+22=63$

EXAMPLE: If the sum of two consecutive even integers is 42, find the numbers.

SOLUTION:

GOAL: You are being asked to find two consecutive even integers whose sum is 42.

STRATEGY: Let $x=$ the first consecutive integer and $x+2=$ the second consecutive integer; the sum is equal to 42. The equation is $x+x+2=42$.

IMPLEMENTATION: Solve the equation:

$$x + x + 2 = 42$$

$$2x + 2 = 42$$

$$2x + 2 - 2 = 42 - 2$$

$$2x = 40$$

$$\frac{2^1 x}{2^1} = \frac{40}{2}$$

$$x = 20 \quad \text{(first even integer)}$$

$$x + 2 = 20 + 2 = 22 \quad \text{(second even integer)}$$

EVALUATION: 20 and 22 are consecutive even integers and their sum is $20 + 22 = 42$.

Try These

1. Three times a number increased by 3 is equal to 30. Find the number.

2. A plumber wishes to cut a 51-inch pipe into three pieces so that each piece is 7 inches longer than the preceding one. Find the length of each piece.

3. The sum of two numbers is 25, and one number is four times the other number. Find the numbers.

4. A basketball team played 20 games and won two more games than it lost. Find the number of games the team won.

5. If one-third of a number is three less than $\frac{1}{2}$ of the number, find the number.

6. If two times a number minus six is equal to 20, find the number.

7. Forty calculators are placed into two boxes so that one box has 4 more calculators than the other box. How many calculators are in each box?

8. If the sum of three consecutive numbers is 39, find the numbers.

9. If the sum of two consecutive odd numbers is 48, find the numbers.

10. Find two numbers whose sum is 30 and whose difference is 6.

SOLUTIONS:

1. Let $x=$ the number; then

$$3x + 3 = 30$$

$$3x + 3 - 3 = 30 - 3$$

$$3x = 27$$

$$\frac{\cancel{3}^1 x}{\cancel{3}^1} = \frac{27}{3}$$

$$x = 9$$

2. Let $x=$ length of first piece; $x+7=$ length of second piece; $x+14=$ length of third piece; then

$$x + x + 7 + x + 14 = 51$$

$$3x + 21 = 51$$

$$3x + 21 - 21 = 51 - 21$$

$$3x = 30$$

$$\frac{\cancel{3}^1 x}{\cancel{3}^1} = \frac{30}{3}$$

$$x = 10 \text{ inches} \quad \text{(first piece)}$$

$$x + 7 = 10 + 7 = 17 \text{ inches} \quad \text{(second piece)}$$

$$x + 14 = 10 + 14 = 24 \text{ inches} \quad \text{(third piece)}$$

3. Let $x =$ one number and $4x =$ the other number; then

$$x + 4x = 25$$

$$5x = 25$$

$$\frac{\cancel{5}^1 x}{\cancel{5}^1} = \frac{25}{5}$$

$$x = 5$$

$$4x = 4 \cdot 5 = 20$$

4. Let $x =$ the number of losses and $x + 2 =$ the number of wins; then

$$x + x + 2 = 20$$

$$2x + 2 = 20$$

$$2x + 2 - 2 = 20 - 2$$

$$2x = 18$$

$$\frac{\cancel{2}^1 x}{\cancel{2}^1} = \frac{18}{2}$$

$$x = 9 \quad \text{(number of losses)}$$

$$x + 2 = 9 + 2 = 11 \quad \text{(number of wins)}$$

5. Let $x =$ the number; then

$$\frac{1}{3}x = \frac{1}{2}x - 3$$

$$\frac{\cancel{6}^2}{1} \cdot \frac{1}{\cancel{3}^1}x = \frac{\cancel{6}^3}{1} \cdot \frac{1}{\cancel{2}^1}x - 6 \cdot 3$$

$$2x = 3x - 18$$

$$2x - 3x = 3x - 3x - 18$$

$$-x = -18$$

$$\frac{-x}{-1} = \frac{-18}{-1}$$

$$x = 18$$

6. Let $x =$ the number; then

$$2x - 6 = 20$$

$$2x - 6 + 6 = 20 + 6$$

$$2x = 26$$

$$\frac{2^1 x}{2^1} = \frac{26}{2}$$

$$x = 13$$

7. Let $x =$ the number of calculators in the first box and $x + 4 =$ the number of calculators in the second box; then

$$x + x + 4 = 40$$

$$2x + 4 = 40$$

$$2x + 4 - 4 = 40 - 4$$

$$2x = 36$$

$$\frac{2^1 x}{2^1} = \frac{36}{2}$$

$$x = 18$$

$$x + 4 = 18 + 4 = 22$$

8. Let $x =$ first integer; $x + 1 =$ second integer; $x + 2 =$ third integer; then

$$x + x + 1 + x + 2 = 39$$

$$3x + 3 = 39$$

$$3x + 3 - 3 = 39 - 3$$

$$3x = 36$$

$$\frac{3^1 x}{3^1} = \frac{36}{3}$$

$$x = 12$$

$$x + 1 = 13$$

$$x + 2 = 14$$

9. Let $x=$ first odd integer and $x+2=$ second odd integer; then

$$x + x + 2 = 48$$

$$2x + 2 = 48$$

$$2x + 2 - 2 = 48 - 2$$

$$2x = 46$$

$$\frac{2^1 x}{2^1} = \frac{46}{2}$$

$$x = 23$$

$$x + 2 = 23 + 2 = 25$$

10. Let $x=$ one number and $30 - x =$ the other number; then

$$x - (30 - x) = 6$$

$$x - 30 + x = 6$$

$$2x - 30 = 6$$

$$2x - 30 + 30 = 30 + 6$$

$$2x = 36$$

$$\frac{2^1 x}{2^1} = \frac{36}{2}$$

$$x = 18$$

$$30 - x = 30 - 18 = 12$$

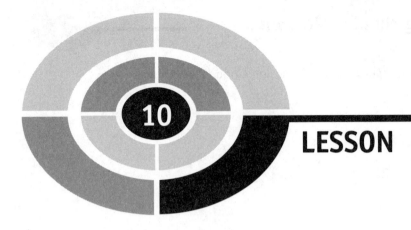

Solving Digit Problems

The symbols 0, 1, 2, 3, 4, 5, 6, 7, 8, and 9 are called **digits**. They are used to make our numbers. A number such as 28 is called a two-digit number. The eight is the one's digit and the two is the ten's digit. The one's digit is also called the unit's digit. The number 28 means the sum of 2 tens and 8 ones and can be written $2 \times 10 + 8 = 20 + 8$ or 28. The number 537 is called a three-digit number. The seven is the one's digit, the three is the ten's digit, and the five is the hundred's digit. It can be written as $5 \times 100 + 3 \times 10 + 7$ or $500 + 30 + 7 = 537$.

A digit problem will sometimes ask you to find the sum of the digits. In order to do this, just add the digits. For example, the sum of the digits of the number 537 is $5 + 3 + 7 = 15$.

Sometimes digit problems will ask you to reverse the digits. If the digits of the number 36 ($3 \times 10 + 6$) are reversed, the new number is 63 ($6 \times 10 + 3$). Using this information and the material in the previous lesson, you will be able to solve digit problems.

LESSON 10   Solving Digit Problems

EXAMPLE: The sum of the digits of a two-digit number is 15. If the digits are reversed, the new number is 9 more than the original number. Find the number.

SOLUTION:

GOAL: You are being asked to find a certain two-digit number.

STRATEGY: Let $x =$ the ten's digit and $15 - x =$ the one's digit. The original number can be written as $10x + (15 - x)$, and the number with the digits reversed can be written as $10(15 - x) + x$. Since the new number is 9 more than the original number, an equation can be written as

new number $=$ original number $+ 9$

$$10(15 - x) + x = 10x + (15 - x) + 9$$

IMPLEMENTATION: Solve the equation:

$$10(15 - x) + x = 10x + (15 - x) + 9$$
$$150 - 10x + x = 10x + 15 - x + 9$$
$$150 - 9x = 9x + 24$$
$$150 - 9x - 9x = 9x - 9x + 24$$
$$150 - 18x = 24$$
$$150 - 150 - 18x = 24 - 150$$
$$-18x = -126$$
$$\frac{-18^1 x}{-18^1} = \frac{-126}{-18}$$
$$x = 7$$

The ten's digit is 7 and the one's digit is $(15 - 7) = 8$. The number, then, is 78.

LESSON 10 Solving Digit Problems

EVALUATION: Take 78 and reverse the digits to get 87. Subtract $87 - 78 = 9$. Hence, the sum of the digits $7 + 8$ is 15 and the difference of the two numbers is 9.

EXAMPLE: The one's digit of a two-digit number is twice the ten's digit. The sum of the digits of the number is 12. Find the number.

SOLUTION:

GOAL: You are being asked to find a two-digit number.

STRATEGY: Let $x =$ the ten's digit and $2x =$ the unit's digit. If the sum of the digits is 12, the equation is $x + 2x = 12$.

IMPLEMENTATION: Solve the equation:

$$x + 2x = 12$$

$$3x = 12$$

$$x = 4$$

$$2x = 2 \times 4 = 8$$

Hence the number is 48.

EVALUATION: The ten's digit is 4 and the one's digit is twice the ten's digit since $2 \times 4 = 8$. The sum of the digits is $4 + 8 = 12$.

EXAMPLE: In a three-digit number, the one's digit is three more than the hundred's digit, and the ten's digit is one more than the hundred's digit. If the sum of the three digits is 10, find the number.

SOLUTION:

GOAL: You are being asked to find a three-digit number.

STRATEGY: Let $x =$ the hundred's digit and $x + 1 =$ the ten's digit and $x + 3 =$ the one's digit. Since the sum of the digits of the number is 10, the equation is $x + x + 1 + x + 3 = 10$.

IMPLEMENTATION: Solve the equation:

$$x + x + 1 + x + 3 = 10$$

$$3x + 4 = 10$$

$$3x + 4 - 4 = 10 - 4$$

$$3x = 6$$

$$\frac{\cancel{3}^1 x}{\cancel{3}^1} = \frac{6}{3}$$

$$x = 2$$

The hundred's digit is 2. The ten's digit is $x + 1$ or $2 + 1 = 3$. The one's digit is $x + 3$ or $2 + 3 = 5$. The number is 235.

EVALUATION: The sum of $2 + 3 + 5 = 10$.

Try These

1. The sum of the digits of a two-digit number is 10. If the ten's digit is 8 more than the one's digit, find the number.

2. The sum of the digits of a two-digit number is 8. If the ten's digit is three times the one's digit, find the number.

3. The sum of the digits of a two-digit number of a marathon runner is 9. If the digits of the number are reversed, the new number is 27 more than the original number. Find the number.

4. The one's digit of a two-digit number is one more than the ten's digit. If the digits of the number are reversed, the new number is 3 less than twice the original number. Find the number.

5. The sum of digits of a number on a football player's uniform is 11. If the digits are reversed, the new number is seven more than twice the original number. Find the number.

6. In a three-digit number, the hundred's digit is twice the ten's digit and the one's digit is 3 less than the hundred's digit. If the sum of the digits of the number is 7, find the number.

7. The sum of the digits of a two-digit number on a stock car is 16. The ten's digit is two more than the one's digit. Find the number.

8. In a two-digit number, the one's digit is three more than the ten's digit. If the digits are reversed, the new number is one less than three times the original number. Find the number.

9. In a three-digit number, the ten's digit and the one's digit are the same and the hundred's digit is three less than the ten's digit. If the sum of the digits is 9, find the number.

10. In a two-digit number, the one's digit is five less than the ten's digit. If the number is equal to 8 times the sum of its digits, find the number.

SOLUTIONS:

1. Let $x =$ the one's digit and $x + 8 =$ the ten's digit; then

$$x + 8 + x = 10$$

$$2x + 8 = 10$$

$$2x + 8 - 8 = 10 - 8$$

$$2x = 2$$

$$x = 1 \quad \text{(one's digit)}$$

$$x + 8 = 9 \quad \text{(ten's digit)}$$

The number is 91.

2. Let $x =$ the one's digit and $3x =$ the ten's digit; then

$$3x + x = 8$$

$$4x = 8$$

$$\frac{\cancel{4}^{1} x}{\cancel{4}^{1}} = \frac{8}{4}$$

$x = 2$ (one's digit)

$3x = 3 \cdot 2 = 6$ (ten's digit)

The number is 62.

3.   Let $x =$ the ten's digit and $9 - x =$ the one's digit; then
     original number is $10x + (9 - x)$; reversed digits number is $10(9 - x) + x$

$$10x + (9 - x) + 27 = 10(9 - x) + x$$

$$10x + 9 - x + 27 = 90 - 10x + x$$

$$9x + 36 = 90 - 9x$$

$$9x + 9x + 36 = 90 - 9x + 9x$$

$$18x + 36 = 90$$

$$18x + 36 - 36 = 90 - 36$$

$$18x = 54$$

$$\frac{\cancel{18}^{1} x}{\cancel{18}^{1}} = \frac{54}{18}$$

$$x = 3$$ (ten's digit)

$$9 - x = 9 - 3 = 6$$ (one's digit)

The number is 36.

4.   Let $x =$ the ten's digit and $x + 1 =$ the one's digit; then
     original number is $10x + (x + 1)$; reversed digits number is $10(x + 1) + x$

$$2(10x + x + 1) - 3 = 10(x + 1) + x$$

$$2(11x + 1) - 3 = 10x + 10 + x$$

$$22x + 2 - 3 = 11x + 10$$

$$22x - 1 = 11x + 10$$

$$22x - 11x - 1 = 11x - 11x + 10$$

$$11x - 1 = 10$$

$$11x - 1 + 1 = 10 + 1$$

$$11x = 11$$

$$\frac{\cancel{11}^1 x}{\cancel{11}^1} = \frac{11}{11}$$

$$x = 1 \quad \text{(ten's digit)}$$

$$x + 1 = 1 + 1 = 2 \quad \text{(one's digit)}$$

The number is 12.

5. Let $x =$ the ten's digit and $11 - x =$ the one's digit; then the original number is $10x + (11 - x)$; the reversed digits number is $10(11 - x) + x$

$$2(10x + 11 - x) + 7 = 10(11 - x) + x$$

$$2(9x + 11) + 7 = 110 - 10x + x$$

$$18x + 22 + 7 = 110 - 10x + x$$

$$18x + 29 = 110 - 9x$$

$$18x + 9x + 29 = 110 - 9x + 9x$$

$$27x + 29 = 110$$

$$27x + 29 - 29 = 110 - 29$$

$$27x = 81$$

$$\frac{\cancel{27}^1 x}{\cancel{27}^1} = \frac{81}{27}$$

$$x = 3 \quad \text{(ten's digit)}$$

$$11 - x = 11 - 3 = 8 \quad \text{(one's digit)}$$

The number is 38.

6. Let $2x =$ the hundred's digit, $x =$ the ten's digit, and $2x - 3 =$ the one's digit; then

$$2x + x + 2x - 3 = 7$$

$$5x - 3 = 7$$

$$5x - 3 + 3 = 7 + 3$$

$$5x = 10$$

$$\frac{\cancel{5}x^1}{\cancel{5}^1} = \frac{10}{5}$$

$$x = 2 \quad \text{(ten's digit)}$$

$$2x = 2 \cdot 2 = 4 \quad \text{(hundred's digit)}$$

$$2x - 3 = 2 \cdot 2 - 3 = 1 \quad \text{(one's digit)}$$

The number is 421.

7. Let $x =$ the one's digit and $x + 2 =$ the ten's digit; then

$$x + 2 + x = 16$$

$$2x + 2 = 16$$

$$2x + 2 - 2 = 16 - 2$$

$$2x = 14$$

$$\frac{\cancel{2}^1 x}{\cancel{2}^1} = \frac{14}{2}$$

$$x = 7 \quad \text{(one's digit)}$$

$$x + 2 = 7 + 2 = 9 \quad \text{(ten's digit)}$$

The number is 97.

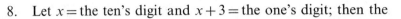
8. Let $x =$ the ten's digit and $x+3 =$ the one's digit; then the original number is $10x + x + 3$; the reversed digits number is $10(x+3) + x$

$$3(10x + x + 3) - 1 = 10(x + 3) + x$$

$$3(11x + 3) - 1 = 10x + 30 + x$$

$$33x + 9 - 1 = 11x + 30$$

$$33x + 8 = 11x + 30$$

$$33x - 11x + 8 = 11x - 11x + 30$$

$$22x + 8 = 30$$

$$22x + 8 - 8 = 30 - 8$$

$$22x = 22$$

$$\frac{\cancel{22}^1 x}{\cancel{22}^1} = \frac{22}{22}$$

$$x = 1 \quad \text{(ten's digit)}$$

$$x + 3 = 1 + 3 = 4 \quad \text{(one's digit)}$$

The number is 14.

9. Let $x =$ the ten's digit, $x =$ the one's digit, and $x - 3 =$ the hundred's digit; then

$$x + x + x - 3 = 9$$

$$3x - 3 = 9$$

$$3x - 3 + 3 = 9 + 3$$

$$3x = 12$$

$$\frac{\cancel{3}^1 x}{\cancel{3}^1} = \frac{12}{3}$$

$x = 4$ (one's and ten's digit)

$x - 3 = 4 - 3 = 1$ (hundred's digit)

The number is 144.

10.   Let $x =$ the ten's digit and $x - 5 =$ the one's digit; then

$$10x + x - 5 = 8(x + x - 5)$$

$$11x - 5 = 8(2x - 5)$$

$$11x - 5 = 16x - 40$$

$$11x - 16x - 5 = 16x - 16x - 40$$

$$-5x - 5 = -40$$

$$-5x - 5 + 5 = -40 + 5$$

$$-5x = -35$$

$$\frac{-\cancel{5}^{1}x}{-\cancel{5}^{1}} = \frac{-35}{-5}$$

$$x = 7 \quad \text{(ten's digit)}$$

$$x - 5 = 7 - 5 = 2 \quad \text{(one's digit)}$$

The number is 72.

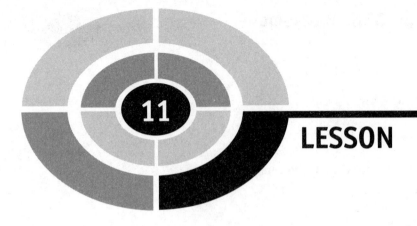

11

Solving Coin Problems

Suppose you have some coins in your wallet. In order to determine the amount of money you have, you would multiply the value of each type of coin by the number of coins of that denomination and then add the answers. For example, if you have 6 nickels, 4 dimes, and 2 quarters, the total amount of money you have in change is

$$6 \times 5\cancel{c} + 4 \times 10\cancel{c} + 2 \times 25\cancel{c}$$
$$= 30\cancel{c} + 40\cancel{c} + 50\cancel{c} = 120\cancel{c} \text{ or } \$1.20.$$

In general, then, to find the amount of money for

Pennies – multiply the number of pennies by $1\cancel{c}$
Nickels – multiply the number of nickels by $5\cancel{c}$
Dimes – multiply the number of dimes by $10\cancel{c}$
Quarters – multiply the number of quarters by $25\cancel{c}$
Half dollars – multiply the number of half dollars by $50\cancel{c}$

In order to avoid decimals, it is easier to work with cents rather than dollars. You can change dollars to cents by multiplying by 100. You can change the answer back to dollars by dividing by 100.

EXAMPLE: A person has 8 coins consisting of quarters and dimes. If the total amount of this change is \$1.25, how many of each kind of coin are there?

SOLUTION:

GOAL: You are being asked to find the number of quarters and the number of dimes the person has.

STRATEGY: Let $x =$ the number of quarters and $(8 - x) =$ the number of dimes; then the value of the quarters is $25x$ and the value of the dimes is $10(8 - x)$. The total amount of money in cents is $\$1.25 \cdot 100 = 125¢$. The equation is $25x + 10(8 - x) = 125$.

IMPLEMENTATION: Solve the equation:

$$25x + 10(8 - x) = 125$$

$$25x + 80 - 10x = 125$$

$$15x + 80 = 125$$

$$15x + 80 - 80 = 125 - 80$$

$$15x = 45$$

$$\frac{\cancel{15}^{1}x}{\cancel{15}^{1}} = \frac{45}{15}$$

$$x = 3 \quad \text{(quarters)}$$

$$8 - x = 8 - 3 = 5 \quad \text{(dimes)}$$

There are 3 quarters and 5 dimes.

EVALUATION: The value of 3 quarters and 5 dimes is $3 \times 25¢ + 5 \times 10¢ = 75¢ + 50¢ = 125¢ = \$1.25$.

EXAMPLE: A person has 3 times as many dimes as he has nickels and 5 more pennies than nickels. If the total amount of these coins is $1.13, how many of each kind of coin does he have?

SOLUTION:

GOAL: You are being asked to find the number of nickels, pennies, and dimes.

STRATEGY: Let $x =$ the number of nickels, $3x =$ the number of dimes, and $x + 5 =$ the number of pennies. Then the value of the nickels is $5x$. The value of the dimes is $10 \cdot 3x$ or $30x$. The value of the pennies is $1 \cdot (x + 5)$. The total amount is 1.13×100 or $113¢$. The equation is $5x + 30x + (x + 5) = 113$.

IMPLEMENTATION: Solve the equation:

$$5x + 30x + (x + 5) = 113$$

$$5x + 30x + x + 5 = 113$$

$$36x + 5 = 113$$

$$36x + 5 - 5 = 113 - 5$$

$$36x = 108$$

$$\frac{36^1 x}{36^1} = \frac{108}{36}$$

$$x = 3 \quad \text{(nickels)}$$

$$3x = 3 \cdot 3 = 9 \quad \text{(dimes)}$$

$$x + 5 = 3 + 5 = 8 \quad \text{(pennies)}$$

There are 3 nickels, 9 dimes, and 8 pennies.

EVALUATION: The value of 3 nickels, 9 dimes, and 8 pennies is $3 \times 5¢ + 9 \times 10¢ + 8 \times 1¢ = 15¢ + 90¢ + 8¢ = 113¢$ or $1.13.

Other types of problems involving values can be solved using the same strategy as the coin problems. Consider the next example.

EXAMPLE: A person bought ten greeting cards consisting of birthday cards costing $1.50 each and anniversary cards costing $2.00 each. If the total cost of the cards was $17.00, find the number of each kind of card the person bought.

SOLUTION:

GOAL: You are being asked to find how many birthday cards and how many anniversary cards the person bought.

STRATEGY: Let $x=$ the number of birthday cards and $(10-x)=$ the number of anniversary cards. Since the birthday cards cost $1.50 each, the value of the birthday cards is $1.5x$, and since the anniversary cards cost $2.00 each, the value of the anniversary cards is $2(10-x)$. The equation is $1.5x + 2(10-x) = 17$.

IMPLEMENTATION: Solve the equation:

$$1.5x + 2(10-x) = 17$$

$$1.5x + 20 - 2x = 17$$

$$20 - 0.5x = 17$$

$$20 - 20 - 0.5x = 17 - 20$$

$$-0.5x = -3$$

$$\frac{-0.5^1 x}{-0.5^1} = \frac{-3}{-0.5}$$

$$x = 6 \quad \text{(birthday cards)}$$

$$(10-x) = 10 - 6 = 4 \quad \text{(anniversary cards)}$$

The person bought 6 birthday cards and 4 anniversary cards.

EVALUATION: 6 birthday cards and 4 anniversary cards cost $6 \times \$1.50 + 4 \times \$2.00 = \$9.00 + \$8.00 = \$17.00$.

Try These

1. If a person has three times as many quarters as dimes and the total amount of money is $5.95, find the number of quarters and dimes.

2. A pile of 18 coins consists of pennies and nickels. If the total amount of the coins is 38¢, find the number of pennies and nickels.

3. A small child has 6 more quarters than nickels. If the total amount of the coins is $3.00, find the number of nickels and quarters the child has.

4. A child's bank contains 32 coins consisting of nickels and quarters. If the total amount of money is $3.80, find the number of nickels and quarters in the bank.

5. A person has twice as many dimes as she has pennies and three more nickels than pennies. If the total amount of the coins is $1.97, find the numbers of each type of coin the person has.

6. In a bank, there are three times as many quarters as half dollars and 6 more dimes than half dollars. If the total amount of the money in the bank is $4.65, find the number of each type of coin in the bank.

7. A person bought 12 stamps consisting of 37¢ stamps and 23¢ stamps. If the cost of the stamps is $3.74, find the number of each type of the stamps purchased.

8. A dairy store sold a total of 80 ice cream sandwiches and ice cream bars. If the sandwiches cost $0.69 each and the bars cost $0.75 each and the store made $58.08, find the number of each sold.

9. An office supply store sells college-ruled notebook paper for $1.59 a ream and wide-ruled notebook paper for $2.29 a ream. If a student purchased 9 reams of notebook paper and paid $15.71, how many reams of each type of paper did the student purchase?

10. A clerk is given $75 in bills to put in a cash drawer at the start of a workday. There are twice as many $1 bills as $5 bills and one less $10 bill than $5 bills. How many of each type of bill are there?

SOLUTIONS:

1. Let $x =$ the number of dimes and $3x =$ the number of quarters; then the value of the dimes is $10x$ and the value of the quarters is $25 \cdot 3x$ or $75x$.

$$10x + 25 \cdot 3x = 5.95 \times 100$$

$$10x + 75x = 595$$

$$85x = 595$$

$$\frac{\cancel{85}^1 x}{\cancel{85}^1} = \frac{595}{85}$$

$$x = 7 \text{ dimes}$$

$$3x = 3 \cdot 7 = 21 \text{ quarters}$$

2. Let $x =$ the number of nickels and $18 - x =$ the number of pennies; then the value of the nickels is $5x$ and the value of the pennies is $1 \cdot (18 - x)$.

$$5x + 18 - x = 38$$

$$4x + 18 = 38$$

$$4x + 18 - 18 = 38 - 18$$

$$4x = 20$$

$$\frac{\cancel{4}^1 x}{\cancel{4}^1} = \frac{20}{4}$$

$$x = 5 \text{ nickels}$$

$$18 - x = 18 - 5 = 13 \text{ pennies}$$

3. Let $x =$ the number of nickels and $x + 6 =$ the number of quarters; then the value of the nickels is $5x$ and the value of the quarters is $25(x + 6)$.

$$5x + 25(x + 6) = 3 \times 100$$

$$5x + 25x + 150 = 300$$

$$30x + 150 = 300$$

$$30x + 150 - 150 = 300 - 150$$

$$30x = 150$$

$$\frac{\cancel{30}^{1}x}{\cancel{30}^{1}} = \frac{150}{30}$$

$$x = 5 \text{ nickels}$$

$$x + 6 = 5 + 6 = 11 \text{ quarters}$$

4. Let $x =$ the number of quarters and $32 - x =$ the number of nickels; then the value of the quarters is $25x$ and the value of the nickels is $5(32 - x)$.

$$25x + 5(32 - x) = 3.80 \times 100$$

$$25x + 160 - 5x = 380$$

$$20x + 160 = 380$$

$$20x + 160 - 160 = 380 - 160$$

$$20x = 220$$

$$\frac{\cancel{20}^{1}x}{\cancel{20}^{1}} = \frac{220}{20}$$

$$x = 11 \text{ quarters}$$

$$32 - x = 32 - 11 = 21 \text{ nickels}$$

5. Let $x =$ the number of pennies, $2x =$ the number of dimes, and $x + 3 =$ the number of nickels; then the value of the pennies is $1x$, the value of the dimes is $10 \cdot 2x$, and the value of the nickels is $5(x + 3)$.

$$x + 10 \cdot 2x + 5(x + 3) = 1.97 \times 100$$

$$x + 20x + 5x + 15 = 197$$

$$26x + 15 = 197$$

$$26x + 15 - 15 = 197 - 15$$

$$26x = 182$$

$$\frac{\cancel{26}^{1}x}{\cancel{26}^{1}} = \frac{182}{26}$$

$$x = 7 \text{ pennies}$$

$$2x = 2 \cdot 7 = 14 \text{ dimes}$$

$$x + 3 = 7 + 3 = 10 \text{ nickels}$$

6. Let $x=$ the number of half dollars, $3x=$ the number of quarters, and $x+6=$ the number of dimes; then the value of the half dollars is $50x$, the value of the quarters is $25 \cdot 3x$, and the value of the dimes is $10(x+6)$.

$$50x + 25 \cdot 3x + 10(x+6) = 4.65 \times 100$$

$$50x + 75x + 10x + 60 = 465$$

$$135x + 60 = 465$$

$$135x + 60 - 60 = 465 - 60$$

$$135x = 405$$

$$\frac{\cancel{135}^{1}x}{\cancel{135}^{1}} = \frac{405}{135}$$

$$x = 3 \text{ half dollars}$$

$$3x = 3 \cdot 3 = 9 \text{ quarters}$$

$$x + 6 = 3 + 6 = 9 \text{ dimes}$$

7. Let $x=$ the number of 37¢ stamps and $(12 - x)=$ the number of 23¢ stamps; then the value of the 37¢ stamps is $37x$ and the value of the 23¢ stamps is $23(12 - x)$.

$$37x + 23(12 - x) = 3.74 \times 100$$

$$37x + 276 - 23x = 374$$

$$14x + 276 = 374$$

$$14x + 276 - 276 = 374 - 276$$

$$14x = 98$$

$$\frac{\cancel{14}^1 x}{\cancel{14}^1} = \frac{98}{14}$$

$$x = 7 \qquad 37¢ \text{ stamps}$$

$$12 - x = 12 - 7 = 5 \qquad 23¢ \text{ stamps}$$

8. Let $x =$ the number of sandwiches and $(80 - x) =$ the number of bars; then the cost of the sandwiches is $69x$ and the cost of the bars is $75(80 - x)$.

$$69x + 75(80 - x) = 5808$$

$$69x + 6000 - 75x = 5808$$

$$6000 - 6x = 5808$$

$$6000 - 6000 - 6x = 5808 - 6000$$

$$-6x = -192$$

$$\frac{-\cancel{6}^1 x}{-\cancel{6}} = \frac{-192}{-6}$$

$$x = 32 \text{ sandwiches}$$

$$80 - x = 80 - 32 = 48 \text{ bars}$$

9. Let $x =$ the number of reams of college-ruled paper and $9 - x =$ the number of reams of wide-ruled paper; then the cost of the college-ruled paper is $159x$ and the cost of the wide-ruled paper is $229(9 - x)$.

$$159x + 229(9 - x) = 1571$$

$$159x + 2061 - 229x = 1571$$

$$-70x + 2061 = 1571$$

$$-70x + 2061 - 2061 = 1571 - 2061$$

$$-70x = -490$$

$$\frac{-\cancel{70}^{1}x}{-\cancel{70}^{1}} = \frac{-490}{-70}$$

$$x = 7 \text{ reams of college-ruled paper}$$

$$9 - x = 9 - 7 = 2 \text{ reams of wide-ruled paper}$$

10. Let $x =$ the number of \$5 bills, $2x =$ the number of \$1 bills, and $x - 1 =$ the number of \$10 bills; then the value of the \$5 bills is $5x$, the value of the \$1 bills is $1 \cdot 2x$, and the value of the \$10 bills is $10(x - 1)$.

$$5x + 2x + 10(x - 1) = 75$$

$$5x + 2x + 10x - 10 = 75$$

$$17x - 10 = 75$$

$$17x - 10 + 10 = 75 + 10$$

$$17x = 85$$

$$\frac{\cancel{17}^{1}x}{\cancel{17}^{1}} = \frac{85}{17}$$

$$x = 5 \text{ five-dollar bills}$$

$$2x = 2 \cdot 5 = 10 \text{ one-dollar bills}$$

$$x - 1 = 5 - 1 = 4 \text{ ten-dollar bills}$$

Quiz 3

1. If one number is 7 times another number and the sum of the numbers is 32, the smaller number is
   (a) 2
   (b) 4
   (c) 3
   (d) 5

2. The sum of two numbers is 42 and one number is 6 more than the other number. The smaller number is
   (a) 24
   (b) 26
   (c) 22
   (d) 18

3. If the sum of 3 consecutive numbers is 60, the largest of the three numbers is
   (a) 21
   (b) 22
   (c) 19
   (d) 18

4. If the sum of two even consecutive numbers is 166, the smaller number is
    (a) 81
    (b) 86
    (c) 80
    (d) 82

5. The sum of the digits of a two-digit number is 11. If the digits are reversed, the new number is 45 more than the old number. Find the number.
    (a) 47
    (b) 38
    (c) 56
    (d) 29

6. The ten's digit is two more than the one's digit of a two-digit number. If 17 is added to the number, the answer is 70. Find the number.
    (a) 64
    (b) 53
    (c) 46
    (d) 35

7. In a two-digit number, the ten's digit is 3 more than the one's digit. If the digits are reversed, the difference between the two numbers is 27. Find the number.
    (a) 52
    (b) 63
    (c) 73
    (d) 85

8. A person has 9 more dimes than nickels. If the total amount of money is $1.20, find the number of dimes the person has.
    (a) 9
    (b) 10
    (c) 11
    (d) 12

9. A person has 20 bills consisting of $1 bills and $2 bills. If the total amount of money the person has is $35, find the number of $2 bills the person has.
   (a) 5
   (b) 8
   (c) 10
   (d) 15

10. A bank contains 8 more pennies than nickels and 3 more dimes than nickels. If the total amount of money in the bank is $3.10, find the number of dimes in the bank.
    (a) 17
    (b) 20
    (c) 25
    (d) 15

Solving Age Problems

When you encounter an age problem, you will often see that the problem gives you information about the age of a person in the future or in the past. For example, if a mother is three times as old as her daughter, their present ages can be represented as

Let $x =$ the daughter's age and

$3x =$ the mother's age

Now if the problem gives you information about their ages, say, 6 years from now, you can represent their future ages as

Let $x + 6 =$ the daughter's future age and

$3x + 6 =$ the mother's future age

Likewise, if the problem gives you some information about the ages, say, 5 years ago, you can represent their past ages as

Let $x - 5$ = the daughter's past age and

$3x - 5$ = the mother's past age

The basic strategy for solving age problems is to represent the present ages of the people, represent the past or future ages of the people, and then set up the equation.

EXAMPLE: A mother is ten times as old as her daughter; in 24 years, she will be twice as old as her daughter. Find their present ages.

SOLUTION:

GOAL: You are being asked to find the present ages of the mother and daughter.

STRATEGY: Let x = the daughter's present age and $10x$ = the mother's present age. In 24 years, their ages will be $x + 24$ = the daughter's age and $10x + 24$ = the mother's age. If the mother will be twice as old as her daughter in 24 years, the equation is 2 times the daughter's age in 24 years = the mother's age in 24 years or $2(x + 24) = 10x + 24$.

IMPLEMENTATION: Solve the equation:

$$2(x + 24) = 10x + 24$$

$$2x + 48 = 10x + 24$$

$$2x - 10x + 48 = 10x - 10x + 24$$

$$-8x + 48 = 24$$

$$-8x + 48 - 48 = 24 - 48$$

$$-8x = -24$$

$$\frac{-8^1 x}{-8^1} = \frac{-24}{-8}$$

$$x = 3 \quad \text{(daughter's age)}$$

$$10x = 10 \cdot 3 = 30 \quad \text{(mother's age)}$$

EVALUATION: In 24 years, the daughter's age is $3 + 24 = 27$ and the mother's age is $30 + 24 = 54$. Since $54 = 2 \cdot 27$, the mother will be twice as old as the daughter.

EXAMPLE: Bill is 8 years older than his brother. In 3 years, Bill will be twice as old as his brother. Find their present ages.

SOLUTION:

GOAL: You are being asked to find the present ages of Bill and his brother.

STRATEGY: Let $x =$ Bill's brother's age and $x + 8 =$ Bill's age. In 3 years, their ages will be $x + 3 =$ Bill's brother's age and $(x + 8) + 3 =$ Bill's age. Now in 3 years, Bill will be twice as old. This means the equation is 2 times Bill's brother's age in 3 years = Bill's age in 3 years or $2(x + 3) = (x + 8) + 3$.

IMPLEMENTATION: Solve the equation:

$$2(x + 3) = (x + 8) + 3$$

$$2x + 6 = x + 8 + 3$$

$$2x + 6 = x + 11$$

$$2x - x + 6 = x - x + 11$$

$$x + 6 = 11$$

$$x + 6 - 6 = 11 - 6$$

$$x = 5 \quad \text{(Bill's brother's age)}$$

$$x + 8 = 5 + 8 = 13 \quad \text{(Bill's age)}$$

EVALUATION: In 3 years, Bill's brother will be $5 + 3 = 8$ years, and Bill will be $13 + 3 = 16$, which is twice his brother's age.

EXAMPLE: Jan is 6 years older than Mary. If the sum of their ages is 32, find each one's age.

SOLUTION:

GOAL: You are being asked to find the ages of Jan and Mary.

STRATEGY: Let $x =$ Mary's age and $x + 6 =$ Jan's age. Then the sum of their ages is $x + x + 6 = 32$.

IMPLEMENTATION: Solve the equation:

$$x + x + 6 = 32$$

$$2x + 6 = 32$$

$$2x + 6 - 6 = 32 - 6$$

$$2x = 26$$

$$\frac{2^1 x}{2^1} = \frac{26}{2}$$

$$x = 13 \quad \text{(Mary's age)}$$

$$x + 6 = 13 + 6 = 19 \quad \text{(Jan's age)}$$

EVALUATION: Jan is 6 years older than Mary and the sum of their ages is $13 + 19 = 32$.

EXAMPLE: A father is 34 years old and his son is 12 years old. In how many years will the father be twice as old as his son?

SOLUTION:

GOAL: You are being asked to find the number of years it will be when the father is twice as old as his son.

STRATEGY: Let $x =$ the number of years. Then the father's age in x years will be $34 + x$ years and the son's age in x years will be $12 + x$ years. If the father is twice as old as the son in x years, the equation is $2(12 + x) = 34 + x$.

IMPLEMENTATION: Solve the equation:

$$2(12 + x) = 34 + x$$

$$24 + 2x = 34 + x$$

$$24 + 2x - x = 34 + x - x$$

$$24 + x = 34$$

$$24 - 24 + x = 34 - 24$$

$$x = 10$$

Hence, in 10 years the father will be twice as old as his son.

EVALUATION: In 10 years, the father will be $34 + 10 = 44$ years old, and the son will be $12 + 10 = 22$ years, in which case the father is twice as old as his son.

Try These

1. A man is six times as old as his son. In 9 years he will be three times as old as his son. How old are they now?

2. A woman is twice as old as her daughter. Twenty years ago, she was four times as old as her daughter. How old are they now?

3. Mark is 4 years older than his brother Mike. If the sum of their ages is 20, how old are they now?

4. Marie is 12 years older than Mary. Nine years ago, Marie was twice as old as Mary. Find their present ages.

5. Sam is 18 and Bill is 24. How many years ago was Bill three times as old as Sam?

6. Pat is five years older than her brother. Two years from now, the sum of their ages will be 23. Find their present ages.

7. The sum of Tyler and Alane's ages is 36. Twelve years ago, Alane was twice as old as Tyler. Find their present ages.

8. Tara is two years older than Ashley. In 4 years from now, Tara will be twice as old as Ashley was 4 years ago. Find their present ages.

9. A father is three times as old as his twin sons. If the sum of their ages in two years will be 81, how old are they now?

10.   Will is 4 years older than Phil. Three years from now, Will will be twice as old as Phil was last year. Find their present ages.

SOLUTIONS:

1.   Let x = the son's age and $6x$ = the father's age; then $x + 9$ = the son's age and $6x + 9$ = the father's age in 9 years.

$$3(x + 9) = 6x + 9$$

$$3x + 27 = 6x + 9$$

$$3x - 6x + 27 = 6x - 6x + 9$$

$$-3x + 27 = 9$$

$$-3x + 27 - 27 = 9 - 27$$

$$-3x = -18$$

$$\frac{-3^1 x}{-3^1} = \frac{-18}{-3}$$

$$x = 6 \quad \text{(son's age)}$$

$$6x = 6 \cdot 6 = 36 \quad \text{(father's age)}$$

2.   Let x = daughter's age and $2x$ = mother's age; then $x - 20$ = the daughter's age and $2x - 20$ = the mother's age 20 years ago.

$$4(x - 20) = 2x - 20$$

$$4x - 80 = 2x - 20$$

$$4x - 2x - 80 = 2x - 2x - 20$$

$$2x - 80 = -20$$

$$2x - 80 + 80 = -20 + 80$$

$$2x = 60$$

$$\frac{2^1 x}{2^1} = \frac{60}{2}$$

$$x = 30 \quad \text{(daughter's age)}$$

$$2x = 2 \cdot 30 = 60 \quad \text{(mother's age)}$$

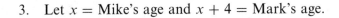

3. Let x = Mike's age and $x + 4$ = Mark's age.

$$x + x + 4 = 20$$

$$2x + 4 = 20$$

$$2x + 4 - 4 = 20 - 4$$

$$2x = 16$$

$$\frac{2^1 x}{2^1} = \frac{16}{2}$$

$$x = 8 \quad \text{(Mike's age)}$$

$$x + 4 = 8 + 4 = 12 \quad \text{(Mark's age)}$$

4. Let x = Mary's age and $x + 12$ = Marie's age; then $x - 9$ = Mary's age and $x + 12 - 9$ = Marie's age 9 years ago.

$$2(x - 9) = x + 12 - 9$$

$$2x - 18 = x + 3$$

$$2x - x - 18 = x - x + 3$$

$$x - 18 = 3$$

$$x - 18 + 18 = 3 + 18$$

$$x = 21 \quad \text{(Mary's age)}$$

$$x + 12 = 21 + 12 = 33 \quad \text{(Marie's age)}$$

5. Let x = number of years ago when Bill was three times as old as Sam; then $18 - x$ = Sam's age and $24 - x$ = Bill's age x years ago.

$$3(18 - x) = 24 - x$$

$$54 - 3x = 24 - x$$

$$54 - 3x + x = 24 - x + x$$

$$54 - 2x = 24$$

$$54 - 54 - 2x = 24 - 54$$

$$-2x = -30$$

$$\frac{-2^1 x}{-2^1} = \frac{-30}{-2}$$

$$x = 15 \text{ years ago}$$

6. Let $x =$ Pat's brother's age and $x + 5 =$ Pat's age; then $x + 2 =$ Pat's brother's age and $x = 5 + 2 =$ Pat's age in two years.

$$x + 2 + x + 5 + 2 = 23$$

$$2x + 9 = 23$$

$$2x + 9 - 9 = 23 - 9$$

$$2x = 14$$

$$\frac{2x}{2} = \frac{14}{2}$$

$$x = 7 \quad \text{(Pat's brother's age)}$$

$$x + 5 = 7 + 5 = 12 \quad \text{(Pat's age)}$$

7. Let $x =$ Tyler's age and $36 - x =$ Alane's age; then $x - 12 =$ Tyler's age and $36 - x - 12 =$ Alane's age 12 years ago.

$$2(x - 12) = 36 - x - 12$$

$$2x - 24 = 24 - x$$

$$2x + x - 24 = 24 - x + x$$

$$3x - 24 = 24$$

$$3x - 24 + 24 = 24 + 24$$

$$3x = 48$$

$$\frac{\cancel{3}^1 x}{\cancel{3}^1} = \frac{48}{3}$$

$$x = 16 \quad \text{(Tyler's age)}$$

$$36 - x = 36 - 16 = 20 \quad \text{(Alane's age)}$$

8. Let x = Ashley's age and $x + 2$ = Tara's age; then $x - 4$ = Ashley's age 4 years ago and $x + 2 + 4$ = Tara's age in 4 years.

$$2(x - 4) = x + 2 + 4$$

$$2x - 8 = x + 6$$

$$2x - x - 8 = x - x + 6$$

$$x - 8 = 6$$

$$x - 8 + 8 = 6 + 8$$

$$x = 14 \quad \text{(Ashley's age)}$$

$$x + 2 = 14 + 2 = 16 \quad \text{(Tara's age)}$$

9. Let x = age of the sons and $3x$ = age of the father; then $x + 2$ = the sons' ages and $3x + 2$ = the father's age in 2 years.

$$x + 2 + x + 2 + 3x + 2 = 81$$

$$5x + 6 = 81$$

$$5x + 6 - 6 = 81 - 6$$

$$5x = 75$$

$$\frac{\cancel{5}^1 x}{\cancel{5}^1} = \frac{75}{5}$$

$$x = 15 \quad \text{(each son's age)}$$

$$3x = 3 \cdot 15 = 45 \quad \text{(father's age)}$$

10. Let $x =$ Phil's age and $x + 4 =$ Will's age; then $x - 1 =$ Phil's age one year ago and $x + 4 + 3 =$ Will's age in 3 years.

$$x + 4 + 3 = 2(x - 1)$$

$$x + 7 = 2x - 2$$

$$x - x + 7 = 2x - x - 2$$

$$7 = x - 2$$

$$7 + 2 = x - 2 + 2$$

$$9 = x \quad \text{(Phil's age)}$$

$$x + 4 = 9 + 4 = 13 \quad \text{(Will's age)}$$

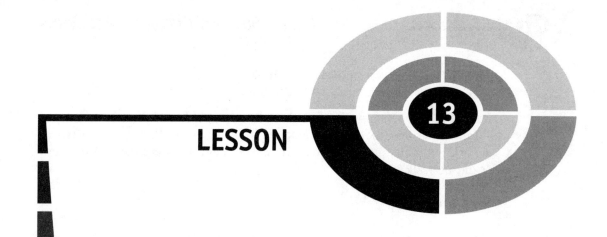

Solving Distance Problems

miles per hour

The basic formula for solving distance problems is distance = rate × time or $D = RT$. For example, if an automobile travels at 30 miles per hour for 2 hours, then the distance is $D = RT$ or $30 \cdot 2 = 60$ miles.

Distance problems usually involve two vehicles (i.e., automobiles, trains, bicycles, etc.) either traveling in the same direction or in opposite directions, or one vehicle making a round trip. The strategy for solving distance problems is

1. Draw a diagram of the situation.
2. Set up a table as shown.

	Rate ×	Time =	Distance
First vehicle			
Second vehicle			

3. Fill in the information in the table.
4. Write an equation for the situation.

EXAMPLE: A person walked from his home to the football stadium at a rate of 3 miles per hour. After the game, he rode the bus back the same way at a rate of 22 miles per hour. If the total time he spent traveling was 2 hours, how far did he walk?

SOLUTION:

GOAL: You are being asked to find the distance the person walked from his home to the stadium.

STRATEGY: The distance he walked and rode is the same, but the direction is different. See Figure 13-1. Place the rate 3 miles per hour for walking and 22 miles per hour for riding in the boxes under Rate. Let $t =$ the time he walked and $2 - t$ the time he rode. Place these in the boxes under Time. To get the distance, multiply the rate by the time and place these expressions in the boxes under Distance as shown.

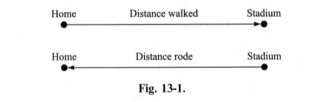

Home   Distance walked   Stadium

Home   Distance rode   Stadium

Fig. 13-1.

	Rate ×	Time =	Distance
Walking	3	t	$3t$
Riding	22	$2 - t$	$22(2 - t)$

Since the distances are equal, the equation is $3t = 22(2 - t)$.

IMPLEMENTATION: Solve the equation:

$$3t = 22(2 - t)$$

$$3t = 44 - 22t$$

$$3t + 22t = 44 - 22t + 22t$$

$$25t = 44$$

$$\frac{25^1 t}{25^1} = \frac{44}{25}$$

$$t = 1.76 \text{ hours}$$

The distance, then, is $D = RT$ or $D = 3 \times 1.76 = 5.28$ miles.

EVALUATION: You can check the answer by determining if the distance the person rode is the same distance he walked.

$$D = RT$$

$$D = 22(2 - 1.76)$$

$$= 22(0.24)$$

$$= 5.28 \text{ miles}$$

In the previous problem, the same person made a round trip. In the next problem, we have two vehicles going in the same direction.

EXAMPLE: A freight train leaves Pittsburgh traveling east at 15 miles per hour. Four hours later, an AMTRAK train leaves Pittsburgh traveling east at 35 miles per hour. How many miles from Pittsburgh will the AMTRAK overtake the freight train?

SOLUTION:

GOAL: You are being asked the distance from Pittsburgh two trains travel before meeting.

STRATEGY: In this case, both trains are traveling in the same direction, and they travel the same distance. See Figure 13-2. Place 35 miles per hour in the top box under Rate and 15 miles per hour in the bottom box under Rate. Then

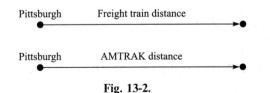

Fig. 13-2.

let t = the time the AMTRAK train travels and place t in the top box under Time. Let $(t+4)$ be the time the freight train travels and place $(t+4)$ in the lower box under time. The freight train travels 4 more hours since it left earlier. To get the distance for each train, multiply the rate by the time and place these expressions under Distance.

	Rate	×	Time	=	Distance
AMTRAK	35		t		$35t$
Freight train	15		$t+4$		$15(t+4)$

Since the distances each train travels are the same, the equation is $35t = 15(t+4)$.

IMPLEMENTATION: Solve the equation:

$$35t = 15(t+4)$$

$$35t = 15t + 15 \cdot 4$$

$$35t = 15t + 60$$

$$35t - 15t = 15t - 15t + 60$$

$$20t = 60$$

$$\frac{\cancel{20}^1 t}{\cancel{20}^1} = \frac{60}{20}$$

$$t = 3 \text{ hours}$$

To find the distance, use the formula $D = RT$.

$$D = RT$$

$$D = 35 \times 3$$

$$= 105 \text{ miles}$$

Hence the AMTRAK train will overtake the freight train 105 miles from Pittsburgh.

EVALUATION: Check to see if the distance the freight train travels is the same as the AMTRAK train; that is, 105 miles.

$$D = RT$$

$$D = 15(3 + 4)$$

$$D = 105 \text{ miles}$$

Another type of distance problem is one where two vehicles are going in the opposite direction.

EXAMPLE: Two automobiles are leaving from the same point and are traveling in opposite directions. One is going 10 miles per hour faster than the other is. After traveling two hours, they are 160 miles apart. How fast is each automobile traveling?

SOLUTION:

GOAL: You are being asked to find the speed in miles per hour that each automobile was traveling.

STRATEGY: Draw a diagram showing that each automobile is traveling in the opposite direction. See Figure 13-3. Let $x =$ the rate (speed) of the first automobile and $x + 10 =$ the rate (speed) of the second automobile. Place these values in the boxes under Rate. Then the time each automobile travels is 2 hours, so place these numbers in the boxes under Time. The expressions that represent the distances are $2x$ and $2(x + 10)$, so place these in the boxes under Distance.

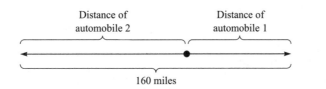

Distance of automobile 2 ⎴   Distance of automobile 1 ⎴

←————————————●————————————→

160 miles

Fig. 13-3.

	Rate	×	Time	=	Distance
First automobile	x		2		$2x$
Second automobile	$x + 10$		2		$2(x + 10)$

If you look at Figure 13-3, you will see that the total distance is the **sum** of the individual distances each car went, since each is going in the opposite direction: $2x + 2(x + 10) = 160$.

IMPLEMENTATION: Solve the equation for x:

$$2x + 2(x + 10) = 160$$
$$2x + 2x + 20 = 160$$
$$4x + 20 = 160$$
$$4x + 20 - 20 = 160 - 20$$
$$4x = 140$$
$$\frac{\cancel{4}^1 x}{\cancel{4}^1} = \frac{140}{4}$$
$$x = 35 \text{ miles per hour}$$
$$x + 10 = 35 + 10 = 45 \text{ miles per hour}$$

Hence the slower automobile was traveling at 35 miles per hour and the faster automobile was traveling at 45 miles per hour.

EVALUATION: We can find the distance each automobile traveled and then see if the sum is 160 miles.

Automobile 1: $D = RT$

                       $D = 35 \times 2 = 70$ miles

Automobile 2: $D = RT$

                       $D = 45 \times 2 = 90$ miles

Hence, $70 + 90 = 160$ miles.

Try These

1. Two people leave from two towns that are 195 miles apart at the same time and travel along the same road toward each other. The first

person drives 5 miles slower than the second person. If they meet in 3 hours, at what rate of speed did each travel?

2. Two planes leave the same airport and travel in opposite directions. Their speeds are 150 miles per hour and 100 miles per hour, respectively. In how many hours will they be 500 miles apart?

3. Two people travel in opposite directions after leaving at the same time from the same place. If one person walks twice as fast as the other, and in two hours they are 10.5 miles apart, find the walking speeds of each.

4. In order to return her friend's bicycle, a girl rides it to her friend's house at a speed of 9 miles per hour. She then walks back home at a speed of 3 miles per hour. If the total time of the round trip was 1.5 hours, how far was her friend's house?

5. A person riding a motorcycle leaves an hour after a person riding a bicycle. Both travel the same road. If the person riding the bicycle is traveling at 10 miles per hour and the person riding the motorcycle is traveling at 30 miles per hour, how long will it take the motorcycle to overtake the bicycle?

SOLUTIONS:

1. Let $x =$ the speed (rate) of the first person and $x + 5 =$ the rate of the second person.

	Rate $\times$	Time $=$	Distance
First person	x	3	$3x$
Second person	$x + 5$	3	$3(x + 5)$

The total distance they travel is 195 miles.

$$3x + 3(x + 5) = 195$$
$$3x + 3x + 15 = 195$$
$$6x + 15 = 195$$
$$6x + 15 - 15 = 195 - 15$$
$$6x = 180$$
$$\frac{\cancel{6}^{1}x}{\cancel{6}^{1}} = \frac{180}{60}$$

$$x = 30 \text{ miles per hour}$$
$$x + 5 = 30 + 5 = 35 \text{ miles per hour}$$

2. Let $t =$ the time of the planes.

	Rate $\times$	Time $=$	Distance
Plane one	150	t	$150t$
Plane two	100	t	$100t$

The total distance is 500 miles since they are going in opposite directions.

$$150t + 100t = 500$$
$$250t = 500$$
$$\frac{250^1 t}{250^1} = \frac{500}{250}$$
$$t = 2 \text{ hours}$$

3. Let $x =$ the speed (rate) of the first person and $2x =$ the rate of the second person.

	Rate $\times$	Time $=$	Distance
First person	x	2	$2x$
Second person	$2x$	2	$2(2x)$

The total distance is 10.5 miles since they are going in opposite directions.

$$2x + 2(2x) = 10.5$$
$$2x + 4x = 10.5$$
$$6x = 10.5$$
$$\frac{6^1 x}{6} = \frac{10.5}{6}$$
$$x = 1.75 \text{ miles per hour}$$
$$2x = 2 \cdot 1.75 = 3.5 \text{ miles per hour}$$

4. Let $t=$ the time it takes to ride to the friend's house and $1.5-t=$ the time it takes to walk back.

	Rate	×	Time	=	Distance
To friend's house	9		t		$9t$
Return home	3		$(1.5-t)$		$3(1.5-t)$

The distances are equal since she is making a round trip.

$$9t = 3(1.5 - t)$$
$$9t = 4.5 - 3t$$
$$9t + 3t = 4.5 - 3t + 3t$$
$$12t = 4.5$$
$$\frac{\cancel{12}^1 t}{\cancel{12}^1} = \frac{4.5}{12}$$
$$t = 0.375 \text{ hour}$$
$$D = RT$$
$$D = 9 \cdot 0.375 = 3.375 \text{ miles}$$

5. Let $t=$ the time the person on the motorcycle takes to overtake the person on the bicycle and $t+1=$ the time the person on the bicycle rides.

	Rate	×	Time	=	Distance
Bicycle	10		$t+1$		$10(t+1)$
Motorcycle	30		t		$30t$

The distances are equal.

$$30t = 10(t + 1)$$
$$30t = 10t + 10$$
$$30t - 10t = 10t - 10t + 10$$
$$20t = 10$$
$$\frac{\cancel{20}^1 t}{\cancel{20}^1} = \frac{10}{20}$$
$$t = 0.5 \text{ hour}$$

LESSON 13 Solving Distance Problems

It will take 0.5 hour or half an hour to overtake the person on the bicycle.

Another type of distance problem involves an airplane flying **with** or **against** the wind or a boat moving **with** or **against** the current. If an airplane is flying in a **headwind**, the speed of the airplane is slowed down by the force of the wind. If an airplane is flying in a **tailwind**, the speed of the airplane is increased by the wind. For example, if an airplane is flying at an airspeed of 150 miles per hour and there is a 30 mile per hour tailwind, then the ground speed of the airplane is actually $150 + 30 = 180$ miles per hour. The airspeed is the speed of the plane as shown on its speedometer, but if you were standing on the ground, you would clock the speed at 180 miles per hour. If the plane had an airspeed of 150 miles per hour and it was flying in a headwind of 30 miles per hour, the ground speed of the airplane would be $150 - 30 = 120$ miles per hour. In order to solve these problems using algebra, the direction of the wind must be parallel to the destination of the airplane. When it is not, trigonometry must be used.

In a similar situation, if a boat is moving downstream at 25 miles per hour (indicated on its speedometer) and the current is 3 miles per hour, then the actual speed of the boat is $25 + 3 = 28$ miles per hour since the current is actually pushing the boat. If the boat is going upstream against the current, then the current is pushing against the boat and holding it back. In this case, the speed of the boat is $25 - 3 = 22$ miles per hour.

EXAMPLE: A boat's speedometer reads 20 miles per hour going downstream and it reaches its destination in $\frac{3}{4}$ of an hour. If the return trip takes one hour at the speed of 20 miles per hour, how fast is the current?

SOLUTION:

GOAL: You are being asked to find the speed (rate) of the current.

STRATEGY: Let $x =$ the rate of the current; then the speed of the boat downstream is $20 + x$ and upstream is $20 - x$. The times are given.

	Rate	×	Time	=	Distance
Downstream	$20 + x$		$\dfrac{3}{4}$		$\dfrac{3}{4}(20 + x)$
Upstream	$20 - x$		1		$1(20 - x)$

Since the distances are equal, the equation is $\frac{3}{4}(20 + x) = 1(20 - x)$.

IMPLEMENTATION: Solve the equation:

$$\frac{3}{4}(20 + x) = 1(20 - x)$$

$$\frac{\cancel{4}^{1}}{1} \cdot \frac{3}{\cancel{4}^{1}}(20 + x) = \frac{4}{1} \cdot (20 - x)$$

$$3(20 + x) = 4(20 - x)$$

$$60 + 3x = 80 - 4x$$

$$60 + 3x + 4x = 80 - 4x + 4x$$

$$60 + 7x = 80$$

$$60 - 60 + 7x = 80 - 60$$

$$7x = 20$$

$$\frac{\cancel{7}^{1}x}{\cancel{7}} = \frac{20}{7}$$

$$x = 2\frac{6}{7} \text{ miles per hour}$$

EVALUATION: Check to see if the distance going downstream is equal to the distance going upstream using $D = RT$.

$$\text{Downstream } D = \frac{3}{4}(20 + x)$$

$$= \frac{3}{4}\left(20 + 2\frac{6}{7}\right)$$

$$= \frac{3}{4}\left(22\frac{6}{7}\right)$$

$$= 17\frac{1}{7} \text{ miles}$$

$$\text{Upstream } D = 1(20 - x)$$

$$= 1\left(20 - 2\frac{6}{7}\right)$$

$$= 17\frac{1}{7} \text{ miles}$$

EXAMPLE: An airplane flies from Pittsburgh to Harrisburg in 1.5 hours and returns in 2 hours. If the wind speed is 16 miles per hour blowing from the west, find the airspeed of the plane.

SOLUTION:

GOAL: You are being asked to find the airspeed of the plane.

STRATEGY: Let $x =$ the airspeed of the plane. Since Harrisburg is east of Pittsburgh and the wind is blowing from west to east, the ground speed from Pittsburgh to Harrisburg is $x + 16$. The ground speed from Harrisburg to Pittsburgh is $x - 16$. The times are given.

	Rate $\times$	Time $=$	Distance
To Harrisburg	$x + 16$	1.5	$1.5(x + 16)$
To Pittsburgh	$x - 16$	2	$2(x - 16)$

Since the distances are the same, the equation is $1.5(x + 16) = 2(x - 16)$.

IMPLEMENTATION: Solve the equation:

$$1.5(x + 16) = 2(x - 16)$$
$$1.5x + 24 = 2x - 32$$
$$1.5x - 2x + 24 = 2x - 2x - 32$$
$$-0.5x + 24 - 24 = -32 - 24$$
$$-0.5x = -56$$
$$\frac{-0.5^1 x}{-0.5^1} = \frac{-56}{-0.5}$$
$$x = 112 \text{ miles per hour}$$

EVALUATION: Check to see if the distances are the same. Use $D = RT$.

$$\text{To Harrisburg:} \quad D = 1.5(x + 16)$$
$$= 1.5(112 + 16)$$
$$= 1.5(128)$$
$$= 192 \text{ miles}$$

To Pittsburgh: $D = 2(x - 16)$
$$= 2(112 - 16)$$
$$= 2(96)$$
$$= 192 \text{ miles}$$

Try These

1. A plane flies from New Eagle to North Oak in 3 hours and returns in 4 hours. If the speed of the wind is 25 miles per hour and it is blowing in the direction of North Oak from New Eagle, find the airspeed of the plane.

2. If a boat travels upstream from point A to point B in $2\frac{1}{2}$ hours and returns downstream from point B to point A in 2 hours, find the speed of the boat (on its speedometer) if the current is 3 miles per hour.

3. If a plane flies from Mount Pleasant to Waltersville in 5 hours with a headwind of 25 miles per hour and returns in 4 hours with a tailwind of 20 miles per hour, find the airspeed of the plane.

4. A boat's speed on its speedometer reads 18 miles per hour going downstream, and it reaches its destination in 1.25 hours. The return trip takes 1.75 hours at 18 miles per hour on the speedometer. Find the speed of the current.

5. A plane flies in a headwind of 22 miles per hour from East Grove to Uniontown in 6 hours and returns in 4 hours with a tailwind of 22 miles per hour. Find the distance between the airports.

SOLUTIONS:

1. Let $x =$ the airspeed of the plane

	Rate	×	Time	=	Distance
To North Oak	$x + 25$		3		$3(x + 25)$
To New Eagle	$x - 25$		4		$4(x - 25)$

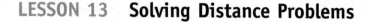
$$3(x + 25) = 4(x - 25)$$
$$3x + 75 = 4x - 100$$
$$3x - 4x + 75 = 4x - 4x - 100$$
$$-x + 75 = -100$$
$$-x + 75 - 75 = -100 - 75$$
$$-x = -175$$
$$\frac{-x}{-1} = \frac{-175}{-1}$$
$$x = 175 \text{ miles per hour}$$

2.   Let $x =$ the speed of the boat as indicated on its speedometer

	Rate	$\times$	Time	$=$	Distance
Upstream	$x - 3$		$2\frac{1}{2}$		$2\frac{1}{2}(x - 3)$
Downstream	$x + 3$		2		$2(x + 3)$

$$2\frac{1}{2}(x - 3) = 2(x + 3)$$
$$\frac{5}{2}(x - 3) = 2(x + 3)$$
$$\frac{\cancel{2}^1}{1} \cdot \frac{5}{\cancel{2}^1}(x - 3) = 2 \cdot 2(x + 3)$$
$$5(x - 3) = 4(x + 3)$$
$$5x - 15 = 4x + 12$$
$$5x - 4x - 15 = 4x - 4x + 12$$
$$x - 15 = 12$$
$$x - 15 + 15 = 12 + 15$$
$$x = 27 \text{ miles per hour}$$

3.   Let $x =$ the airspeed of the plane

	Rate	$\times$	Time	$=$	Distance
To Waltersville	$x - 25$		5		$5(x - 25)$
To Mt. Pleasant	$x + 20$		4		$4(x + 20)$

$$5(x - 25) = 4(x + 20)$$

$$5x - 125 = 4x + 80$$

$$5x - 4x - 125 = 4x - 4x + 80$$

$$x - 125 = 80$$

$$x - 125 + 125 = 80 + 125$$

$$x = 205 \text{ miles per hour}$$

4. Let $x =$ the speed of the current

	Rate	× Time	= Distance
Downstream	$18 - x$	1.75	$1.75(18 - x)$
Upstream	$18 + x$	1.25	$1.25(18 + x)$

$$1.25(18 + x) = 1.75(18 - x)$$

$$22.5 + 1.25x = 31.5 - 1.75x$$

$$22.5 + 1.25x + 1.75x = 31.5 - 1.75x + 1.75x$$

$$22.5 + 3x = 31.5$$

$$22.5 - 22.5 + 3x = 31.5 - 22.5$$

$$3x = 9$$

$$\frac{\cancel{3}^1 x}{\cancel{3}^1} = \frac{9}{3}$$

$$x = 3 \text{ miles per hour}$$

5. Let $x =$ the airspeed of the plane

	Rate	× Time	= Distance
To East Grove	$x - 22$	6	$6(x - 22)$
To Uniontown	$x + 22$	4	$4(x + 22)$

$$4(x + 22) = 6(x - 22)$$
$$4x + 88 = 6x - 132$$
$$4x - 6x + 88 = 6x - 6x - 132$$
$$-2x + 88 = -132$$
$$-2x + 88 - 88 = -132 - 88$$
$$-2x = -220$$
$$\frac{-\cancel{2}^{1} x}{-\cancel{2}^{1}} = \frac{-220}{-2}$$
$$x = 110 \text{ miles per hour}$$
$$D = RT = 4(110 + 22) = 528 \text{ miles}$$

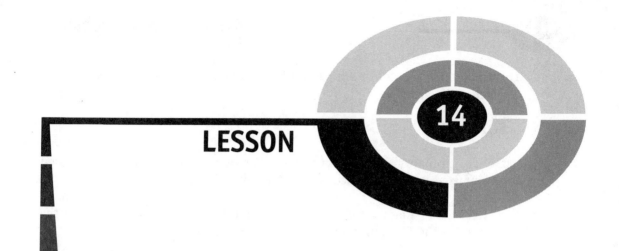

LESSON 14

Solving Mixture Problems

Many real-life problems involve mixtures. There are two basic types of mixture problems. One type uses percents. For example, a metal worker may wish to combine two alloys of different percentages of copper to make a third alloy consisting of a specific percentage of copper. In this case, it is necessary to remember that the percent of the specific substance in the mixture times the amount of mixture is equal to the amount of the pure substance in the mixture. Suppose you have 64 ounces of a mixture consisting of alcohol and water and 30% of it is alcohol; then 30% of 64 ounces or 19.2 ounces of the mixture is alcohol.

A table can be used to solve the percent mixture problems and an equation can be written using

Mixture 1 + Mixture 2 = Mixture 3

EXAMPLE: A pharmacist has two bottles of alcohol; one bottle contains a 10% solution of alcohol and the other bottle contains a 5% solution of alcohol. How much of each should be mixed to get 20 ounces of a solution which is 8% alcohol?

SOLUTION:

GOAL: You are being asked to find the amounts of each solution that need to be mixed to get 20 ounces of an 8% solution.

STRATEGY: Let $x=$ the amount of the 10% solution and $20-x=$ the amount of the 5% solution; then set up a table as follows:

	Amount	×	Percent	=	Amount of pure
Mixture 1	x		10%		$10\%x$
Mixture 2	$20-x$		5%		$5\%(20-x)$
Mixture 3	20		8%		$8\%(20)$

The equation is

Mixture 1   +   Mixture 2   = Mixture 3

$10\%x$ + $5\%(20-x)$ = $8\%(20)$

IMPLEMENTATION: Solve the equation:

$$10\%x+5\%(20-x)=8\%(20)$$

Change the percents to decimals before solving the equation.

$$0.10x+0.05(20-x)=0.08(20)$$

$$0.10x+1-0.05x=1.6$$

$$0.05x+1=1.6$$

$$0.05x+1-1=1.6-1$$

$$0.05x=0.6$$

$$\frac{\cancel{0.05}^{1}x}{\cancel{0.05}^{1}} = \frac{0.6}{0.05}$$

$x = 12$ ounces of Mixture 1

$20 - x = 20 - 12 = 8$ ounces of Mixture 2

Hence 12 ounces of the 10% solution should be mixed with 8 ounces of the 5% solution to get 20 ounces of an 8% solution.

EVALUATION: Check the equation:

$$10\%x + 5\%(20 - x) = 8\%(20)$$

$$10\%(12) + 5\%(8) = 8\%(20)$$

$$1.2 + 0.4 = 1.6$$

$$1.6 = 1.6$$

EXAMPLE: A craftsperson has two alloys of silver. The first one is 70% pure silver and the second one is 50% silver. How many ounces of each must be mixed to have 12 ounces of an alloy which is 65% silver?

SOLUTION:

GOAL: You are being asked to find how much of each alloy should be mixed to get 12 ounces of an alloy which is 65% silver.

STRATEGY: Let $x =$ the amount of the 70% silver alloy and $12 - x =$ the amount of the 50% silver alloy; then

	Amount $\times$	Percent $=$	Amount of pure
Alloy 1	x	70%	$70\%x$
Alloy 2	$12 - x$	50%	$50\%(12 - x)$
Alloy 3	12	65%	$65\%(12)$

The equation is

$$\text{Alloy 1} + \text{Alloy 2} = \text{Alloy 3}$$
$$70\%x + 50\%(12 - x) = 65\%(12)$$

IMPLEMENTATION: Solve the equation:

$$70\%x + 50\%(12 - x) = 65\%(12)$$

$$0.70x + 0.50(12 - x) = 0.65(12)$$

$$0.70x + 6 - 0.50x = 7.8$$

$$0.20x + 6 - 6 = 7.8 - 6$$

$$0.2x = 1.8$$

$$\frac{\cancel{0.2}^{1}x}{\cancel{0.2}^{1}} = \frac{1.8}{0.2}$$

$$x = 9 \text{ ounces of Alloy 1}$$

$$12 - x = 12 - 9 = 3 \text{ ounces of Alloy 2}$$

Hence 9 ounces of the 70% alloy should be mixed with 3 ounces of the 50% alloy to get 12 ounces of an alloy which is 65% silver.

EVALUATION: Check the equation:

$$70\%x + 50\%(12 - x) = 65\%(12)$$

$$70\%(9) + 50\%(3) = 65\%(12)$$

$$6.3 + 1.5 = 7.8$$

$$7.8 = 7.8$$

Another type of percent mixture requires that a mixture be **diluted** to make a **weaker** concentration of the mixture. If the amounts are the same, then a portion of the higher concentration mixture must be removed, and the same amount of the weaker mixture must be added. In this case, the equation would look like this:

Mixture 1 − Amount to be + Amount of weaker mixture = Mixture 2
　　　　　　　removed　　　　　　　to be added

The amount of Mixture 1 equals the amount of Mixture 2. The amount to be removed equals the amount to be added.

EXAMPLE: How much antifreeze which is 30% alcohol must be removed from a 48-ounce container and replaced with water to make 48 ounces of a solution which is 20% alcohol?

SOLUTION:

GOAL: You are being asked to find how much of the alcohol solution must be removed and how much water must be added to dilute the solution.

STRATEGY: Let $x=$ the amount of the alcohol solution to be removed and the amount of water to be added; then

Mixture 1 − Amount to be + Amount of weaker mixture = Mixture 2
　　　　　　　removed 　　　　　　 to be added

$$30\%(48) - 30\%(x) + 0\%(x)^* = 20\%(48)$$

*Since the water has no alcohol, 0% is used.

IMPLEMENTATION: Solve the equation:

$$30\%(48) - 30\%x + 0\%(x) = 20\%(48)$$

$$0.30(48) - 0.30\%x = 0.20(48)$$

$$14.4 - 0.3x = 9.6$$

$$14.4 - 14.4 - 0.30x = 9.6 - 14.4$$

$$-0.30x = -4.8$$

$$\frac{-0.30^1 x}{-0.30^1} = \frac{-4.8}{-0.30}$$

$$x = 16 \text{ ounces}$$

Hence to dilute the 30% solution, 16 ounces must be removed and 16 ounces of water must be added to make 48 ounces of a 20% solution.

EVALUATION: Check the equation:

$$30\%(48) - 30\%(x) + 0\%x = 20\%(48)$$

$$30\%(48) - 30\%(16) + 0\%(16) = 20\%(48)$$

$$14.4 - 4.8 = 9.6$$

$$9.6 = 9.6$$

The second type of mixture problems consists of mixing two items such as coffees, teas, candy, etc., with different prices. These problems are similar to the previous ones. You can use this basic equation:

Item 1 × Its price + Item 2 × Its price = Mixture × Its price

EXAMPLE: A merchant mixes some coffee costing $4 a pound with some coffee costing $3 a pound. How much of each must be used in order to make 20 pounds of mixture costing $3.75 per pound?

SOLUTION:

GOAL: You are being asked to find how much of each coffee must be mixed together to get 20 pounds of coffee costing $3.75.

STRATEGY: Let $x =$ the amount of the $4 coffee and $20 - x =$ the amount of the $3 coffee; then

	Amount	×	Price	=	Total value
Coffee 1	x		$4		$4(x)$
Coffee 2	$20 - x$		$3		$3(20 - x)$
Mixture	20		$3.75		$3.75(20)$

The equation is $4x + 3(20 - x) = 3.75(20)$.

IMPLEMENTATION: Solve the equation:

$$4x + 3(20 - x) = 3.75(20)$$

$$4x + 60 - 3x = 75$$

$$x + 60 = 75$$

$$x + 60 - 60 = 75 - 60$$

$$x = 15 \text{ pounds of \$4 coffee}$$

$$20 - x = 20 - 15 = 5 \text{ pounds of \$3 coffee}$$

Hence 15 pounds of coffee costing $4 a pound must be mixed with 5 pounds of coffee costing $3 a pound to get 20 pounds of coffee costing $3.75 a pound.

EVALUATION: Check the equation:

$$4x + 3(20 - x) = 3.75(20)$$

$$4(15) + 3(5) = 3.75(20)$$

$$60 + 15 = 75$$

Try These

1. How much cream that is 20% butterfat should be mixed with milk that is 5% butterfat to get 10 gallons of cream that is 14% butterfat?

2. How much of a 90% alloy must be combined with a 70% gold alloy in order to make 60 ounces of an 85% gold alloy?

3. How much of an alloy that is 40% zinc should be added to 75 pounds of an alloy that is 65% zinc to get an alloy that is 50% zinc?

4. How much of a solution that is 18% fertilizer must be mixed with a solution that is 30% fertilizer to get 50 gallons of a solution that is 27% fertilizer?

5. How much pure alcohol (100%) should be added to 40 ounces of a solution which is 20% alcohol to get 40 ounces of a solution which is 25% alcohol?

6. A merchant blends a $2.50 per pound tea with a $3.75 per pound tea to get 10 pounds of tea costing $3.00 per pound. How many pounds of each did the merchant use?

7. A candy maker mixes some candy costing $1.25 a pound with some candy costing $0.75 per pound to get 5 pounds of candy costing $0.90 per pound. How many pounds of each did he use?

8. A baker has 8 pounds of cookies costing $1.50 per pound and wants to mix them with some cookies costing $2.00 per pound. How much of the latter should be mixed in order to get a mixture of the two types of cookies costing $1.80 per pound?

9. How many pounds of cashew nuts costing $5.00 a pound must be mixed with peanuts costing $2.20 a pound to get 20 pounds of nuts costing $3.18 a pound?

10. A candy store owner wants to make 30 one-pound boxes of candy costing $5.00 a box. If she wishes to use 12 pounds of candy costing $8.00 per pound, what should be the cost of the other type of candy she should use?

SOLUTIONS:

1. Let $x =$ the amount of liquid that is 20% butterfat and $10 - x =$ the amount of liquid that is 5% butterfat.

$$20\%x + 5\%(10 - x) = 14\%(10)$$

$$0.20x + 0.05(10 - x) = 0.14(10)$$

$$0.20x + 0.5 - 0.05x = 1.4$$

$$0.15x + 0.5 - 0.5 = 1.4 - 0.5$$

$$0.15x = 0.9$$

$$\frac{0.\cancel{15}^1 x}{0.\cancel{15}^1} = \frac{0.9}{0.15}$$

$x = 6$ ounces of 20% butterfat liquid

$10 - x = 10 - 6 = 4$ ounces of 5% butterfat liquid

2. Let $x =$ the amount of 90% alloy and $60 - x =$ the amount of 70% alloy.

$$90\%x + 70\%(60 - x) = 85\%(60)$$

$$0.90x + 0.70(60 - x) = 0.85(60)$$

$$0.90x + 42 - 0.7x = 51$$

$$0.20x + 42 - 42 = 51 - 42$$

$$0.20x = 9$$

$$\frac{0.\cancel{20}^1 x}{0.\cancel{20}^1} = \frac{9}{0.20}$$

$x = 45$ ounces of the 90% alloy

$60 - x = 60 - 45 = 15$ ounces of the 70% alloy

3. Let $x =$ the amount of 40% alloy.

$$40\%x + 65\%(75) = 50\%(75 + x)$$

$$0.40x + 0.65(75) = 0.50(75 + x)$$

$$0.40x + 48.75 = 37.5 + 0.50x$$

$$0.40x - 0.50x + 48.75 = 37.5 + 0.50x - 0.50x$$

$$-0.10x + 48.75 = 37.5$$

$$-0.10x + 48.75 - 48.75 = 37.5 - 48.75$$

$$-0.10x = -11.25$$

$$\frac{-0.10^1 x}{-0.10^1} = \frac{-11.25}{-0.10}$$

$$x = 112.5 \text{ pounds}$$

4. Let $x =$ the amount of 18% fertilizer and $50 - x =$ the amount of the 30% fertilizer.

$$18\%x + 30\%(50 - x) = 27\%(50)$$

$$0.18x + 0.30(50 - x) = 0.27(50)$$

$$0.18x + 15 - 0.30x = 13.5$$

$$-0.12x + 15 = 13.5$$

$$-0.12x + 15 - 15 = 13.5 - 15$$

$$-0.12x = -1.5$$

$$\frac{-0.12^1 x}{-0.12^1} = \frac{-1.5}{-0.12}$$

$$x = 12.5 \text{ gallons of the 18\% solution}$$

$$50 - x = 50 - 12.5 = 37.5 \text{ gallons of the 30\% solution}$$

5. Let $x =$ the amount of pure alcohol to be added and the amount of the 20% solution to be removed.

$$20\%(40) - 20\%x + 100\%x = 25\%(40)$$

$$0.20(40) - 0.20x + 1.00x = 0.25(40)$$

$$8 + 0.8x = 10$$

$$8 - 8 + 0.8x = 10 - 8$$

$$0.8x = 2$$

$$\frac{0.8^1 x}{0.8^1} = \frac{2}{0.8}$$

$$x = 2.5 \text{ ounces}$$

Hence 2.5 ounces of pure alcohol must be added to increase the concentration after 2.5 ounces have been removed.

6. Let $x=$ the number of pounds of \$2.50 tea and $10-x=$ the number of pounds of \$3.75 tea.

$$2.50x + 3.75(10 - x) = \$3.00(10)$$

$$2.50x + 37.5 - 3.75x = 30$$

$$-1.25x + 37.5 = 30$$

$$-1.25x + 37.5 - 37.5 = 30 - 37.5$$

$$-1.25x = -7.5$$

$$\frac{-1.25^1 x}{-1.25^1} = \frac{-7.5}{-1.25}$$

$$x = 6 \text{ pounds of \$2.50 tea}$$

$$10 - x = 10 - 6 = 4 \text{ pounds of \$3.75 tea}$$

7. Let $x=$ the number of pounds of candy costing \$1.25 per pound and $5-x=$ the number of pounds of candy costing \$0.75 per pound.

$$1.25x + 0.75(5 - x) = 0.90(5)$$

$$1.25x + 3.75 - 0.75x = 4.50$$

$$0.5x + 3.75 = 4.50$$

$$0.5x + 3.75 - 3.75 = 4.50 - 3.75$$

$$0.5x = 0.75$$

$$\frac{0.5^1 x}{0.5^1} = \frac{0.75}{0.5}$$

$$x = 1.5 \text{ pounds of \$1.25 candy}$$

$$5 - x = 5 - 1.5 = 3.5 \text{ pounds of \$0.75 candy}$$

8.   Let $x =$ the number of pounds of the $2.00 cookies.

$$1.50(8) + 2.00x = 1.80(8 + x)$$

$$12 + 2.00x = 14.4 + 1.8x$$

$$12 + 2.00x - 1.8x = 14.4 + 1.8x - 1.8x$$

$$12 + 0.2x = 14.4$$

$$12 - 12 + 0.2x = 14.4 - 12$$

$$0.2x = 2.4$$

$$\frac{\cancel{0.2}^{1} x}{\cancel{0.2}^{1}} = \frac{2.4}{0.2}$$

$$x = 12 \text{ pounds}$$

9.   Let $x =$ the number of pounds of cashews and $20 - x =$ the number of pounds of peanuts.

$$5.00x + 2.20(20 - x) = 3.18(20)$$

$$5.00x + 44 - 2.20x = 63.6$$

$$2.8x + 44 = 63.6$$

$$2.8x + 44 - 44 = 63.6 - 44$$

$$2.8x = 19.6$$

$$\frac{\cancel{2.8}^{1} x}{\cancel{2.8}^{1}} = \frac{19.6}{2.8}$$

$$x = 7 \text{ pounds of cashews}$$

$$20 - x = 20 - 7 = 13 \text{ pounds of peanuts}$$

10.   Let $x =$ the cost of the other type of candy she will need to use.

$$8.00(12) + x(18) = 5.00(30)$$

$$96 + 18x = 150$$

$$96 - 96 + 18x = 150 - 96$$

$$18x = 54$$

$$\frac{\cancel{18}^{1}x}{\cancel{18}} = \frac{54}{18}$$

$$x = 3 = \$3.00$$

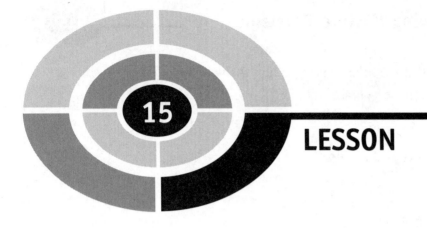

Solving Finance Problems

Finance problems use the basic concepts of investment. There are three terms that are used. The **interest**, also called the **return**, is the amount of money that is made on an investment. The **principal** is the amount of money invested, and the **rate** or **interest rate** is a percent that is used to compute the interest. The basic formula is

Interest = Principal × Rate × Time
or
$I = PRT$

In these problems, the interest used is simple interest per year, and the time is one year. The problems can be set up using a table similar to the ones used in the previous lessons. The equation is derived from the following:

Interest from first + Interest from second = Total interest
    investment   investment

EXAMPLE: A person has $5000 to invest and decides to invest part of it at 4% and the rest of it at $6\frac{1}{2}\%$. If the total interest for the year from the amounts is $245, how much does the person have invested at each rate?

SOLUTION:

GOAL: You are being asked to find the amounts of money invested at 4% and $6\frac{1}{2}\%$.

STRATEGY: Let $x =$ the amount of money invested at 4% and ($5000 - x$) = the amount of money invested at $6\frac{1}{2}\%$. Then set up a table as shown.

	Principal $\times$	Rate $=$	Interest
First investment	x	4%	$4\%x$
Second investment	$5000 - x$	$6\frac{1}{2}\%$	$6\frac{1}{2}\%(5000 - x)$

The equation is

Interest on the first + Interest on second = Total interest
    investment   investment

$$4\%x + 6\frac{1}{2}\%(5000 - x) = \$245$$

IMPLEMENTATION: Solve the equation:

$$4\%x + 6\frac{1}{2}\%(5000 - x) = 245$$

$$0.04x + 0.065(5000 - x) = 245$$

$$0.04x + 325 - 0.065x = 245$$

$$-0.025x + 325 = 245$$

$$-0.025x + 325 - 325 = 245 - 325$$

$$-0.025x = -80$$

$$\frac{-0.025^1 x}{-0.025^1} = \frac{-80}{-0.025}$$

$$x = \$3200 \text{ invested at } 4\%$$

$$5000 - x = 5000 - 3200 = \$1800 \text{ invested at } 6\frac{1}{2}\%$$

EVALUATION: Find the interest on both investments separately and then add them up to see if they equal $245. Use $I = PRT$ where $T = 1$.

First investment: $I = \$3200(4\%) = \$128$

Second investment: $I = \$1800(6\frac{1}{2}\%) = \$117$

$\$128 + \$117 = \$245$

EXAMPLE: A person has twice as much money invested at 6% as he has at 3%. If the total annual interest from the investments is $315, how much does he have invested at each rate?

SOLUTION:

GOAL: You are being asked to find how much money is invested at 6% and 3%.

STRATEGY: Let $x =$ the amount of money invested at 3% and $2x =$ the amount of money invested at 6%; then

	Principal $\times$	Rate $=$	Interest
First investment	$2x$	6%	6%(2x)
Second investment	x	3%	3%(x)

The equation is $6\%(2x) + 3\%(x) = \$315$.

IMPLEMENTATION: Solve the equation:

$$0.06(2x) + 0.03(x) = 315$$

$$0.12x + 0.03x = 315$$

$$0.15x = 315$$

$$\frac{\cancel{0.15}^1 x}{\cancel{0.15}^1} = \frac{315}{0.15}$$

$$x = 2100$$

$$2x = 2(2100) = 4200$$

Hence the person has $2100 invested at 3% and $4200 invested at 6%.

EVALUATION: Find the interest earned on each investment, then add, and see if the sum is $315. Use $I = PRT$ where $T = 1$.

First investment: $I = 3\%(\$2100) = \$63$

Second investment: $I = 6\%(4200) = \$252$

$\$63 + \$252 = \$315$

EXAMPLE: An investor has $800 more invested in stocks paying 5% than she has invested in bonds paying 4%. If the total interest is $103, find the amount of money invested in each.

SOLUTION:

GOAL: You are being asked to find the amount of each investment.

STRATEGY: Let $x =$ the amount invested in the bonds and $x + 800 =$ the amount invested in stocks.

	Principal $\times$	Rate $=$	Interest
Bonds	x	4%	$4\%(x)$
Stocks	$x + 800$	5%	$5\%(x + 800)$

The equation is $4\%x + 5\%(x + 800) = \$103$.

IMPLEMENTATION: Solve the equation:

$$4\%x + 5\%(x + 800) = \$103$$

$$0.04x + 0.05(x + 800) = 103$$

$$0.04x + 0.05x + 40 = 103$$

$$0.09x + 40 = 103$$

$$0.09x + 40 - 40 = 103 - 40$$

$$0.09x = 63$$

$$\frac{\cancel{0.09}^{1}x}{\cancel{0.09}^{1}} = \frac{63}{0.09}$$

$$x = \$700$$

$$x + 800 = 700 + 800 = \$1500$$

Hence the person has $700 invested in bonds and $1500 in stocks.

EVALUATION: Find the interest of both investments and then add to see if the answer is $103. Use $I = PRT$ where $T = 1$.

Bonds: $I = 4\%(700) = \$28$

Stocks: $I = 5\%(1500) = \$75$

$\$28 + \$75 = \$103$

EXAMPLE: An investor has 3 times as much money invested at 5% as he has invested at 2% and $600 more invested at 3% than he has invested at 2%. If the total interest from the three investments is $98, find the amounts he has invested at each rate.

SOLUTION:

GOAL: You are being asked to find the amounts of the three investments.

STRATEGY: Let $x=$ the amount invested at 2%, $3x=$ the amount of money invested at 5%, and $x+600=$ the amount of money invested at 3%.

	Principal ×	Rate =	Interest
First investment	x	2%	2%x
Second investment	$3x$	5%	5%($3x$)
Third investment	$x+600$	3%	3%($x+600$)

The equation is $2\%x + 5\%(3x) + 3\%(x+600) = \$98$

IMPLEMENTATION: Solve the equation:

$$2\%x + 5\%(3x) + 3\%(x+600) = \$98$$

$$0.02x + 0.05(3x) + 0.03(x+600) = 98$$

$$0.02x + 0.15x + 0.03x + 18 = 98$$

$$0.20x + 18 = 98$$

$$0.20x + 18 - 18 = 98 - 18$$

$$0.20x = 80$$

$$\frac{\cancel{0.20}^{1}x}{\cancel{0.20}^{1}} = \frac{80}{0.2}$$

$$x = 400$$

$$3x = 3(400) = 1200$$

$$x + 600 = 400 + 600 = 1000$$

Hence the investor invested $400 at 2%, $1200 at 5%, and $1000 at 3%.

EVALUATION: Find the three interests and add to see if you get $98. Use $I = PRT$ where $T = 1$.

First investment: $I = 2\%(400) = \$8$

Second investment: $I = 5\%(1200) = \$60$

Third investment: $I = 3\%(1000) = \$30$

$\$8 + \$60 + \$30 = \$98$

Try These

1. An investor made two investments, one paying 7% and one paying 5%. If the total amount invested was $12,000 and the total interest she earned after one year was $760, find the amount of each investment.

2. A person invested $9000, part at $4\frac{1}{2}\%$ and the rest at 5%. If the total interest he earned after one year was $435, find the amount of each investment.

3. A person invested twice as much in stocks paying 3% as he did in bonds paying 6%. If the interest at the end of the year was $780, find the amount of money he invested in each.

4. A person invested a certain amount of money in a savings account paying 2% and $1600 more than that amount in a one-year CD paying $3\frac{1}{2}\%$. If the total interest in the two investments was $276, find the amount of money she invested in each.

5. An investor invested a certain amount of money in an account paying 3%. He invests 4 times that amount into another account paying $4\frac{1}{2}\%$, and he invests $900 more than the amount in the 3% account into a third account paying 6%. If the total yearly interest from all three accounts was $297, find the amount he invested in each account.

6. A person invested 5 times as much money at $5\frac{1}{2}\%$ as she did at $2\frac{3}{4}\%$. If the total interest from the investment at the end of the year was $181.50, how much did she invest at each rate?

7. A person has $1300 less invested at 9% than he has invested at 5%. If the total yearly interest from the investments is $219, find the amounts he has invested at each rate.

8. An investor invested some money at 7% and some money at 5%. If the yearly interest on both investments is the same and the total amount of the investments is $12,000, find the amount of each investment.

9. Mr. Jordan has three investments. He has three times the amount of money invested at 4% as he has invested at 1% and $400 more invested at 3% as he has at 1%. If the yearly interest is $108, find the amount of each investment.

10. A person has $3000 invested at 6%. How much should he invest at 4% to have an income (yearly) interest of $372?

SOLUTIONS:

1. Let $x=$ the amount of money invested at 7% and $\$12000 - x=$ the amount of money invested at 5%.

$$7\%x + 5\%(12,000 - x) = 760$$

$$0.07x + 0.05(12,000 - x) = 760$$

$$0.07x + 600 - 0.05x = 760$$

$$0.02x + 600 = 760$$

$$0.02x + 600 - 600 = 760 - 600$$

$$0.02x = 160$$

$$\frac{0.02^1 x}{0.02^1} = \frac{160}{0.02}$$

$$x = \$8000 \text{ at } 7\%$$

$$12,000 - x = 12,000 - 8000 = \$4000 \text{ at } 5\%$$

2. Let $x=$ the amount of money invested at $4\frac{1}{2}\%$ and $\$9000 - x=$ the amount of money invested at 5%.

$$4\frac{1}{2}\%x + 5\%(9000 - x) = \$435$$

$$0.045x + 0.05(9000 - x) = 435$$

$$0.045x + 450 - 0.05x = 435$$

$$-0.005x + 450 = 435$$

$$-0.005x + 450 - 450 = 435 - 450$$

$$-0.005x = -15$$

$$\frac{-0.005^1 x}{-0.005^1} = \frac{-15}{-0.005}$$

$$x = \$3000 \text{ invested at } 4\frac{1}{2}\%$$

$$9000 - x = 9000 - 3000 = \$6000 \text{ invested at } 5\%$$

3. Let $x =$ the amount of money invested at 6% and $2x =$ the amount of money invested at 3%.

$$6\%x + 3\%(2x) = \$780$$

$$0.06x + 0.03(2x) = 780$$

$$0.06x + 0.06x = 780$$

$$0.12x = 780$$

$$\frac{0.12^1 x}{0.12^1} = \frac{780}{0.12}$$

$$x = \$6500 \text{ invested at } 6\%$$

$$2x = 2(6500) = \$13,000 \text{ invested at } 3\%$$

4. Let $x =$ the amount of money invested at 2% and $x + 1600$ the amount of money invested at $3\frac{1}{2}\%$.

$$2\%x + 3\frac{1}{2}\%(x + 1600) = 276$$

$$0.02x + 0.035(x + 1600) = 276$$

$$0.02x + 0.035x + 56 = 276$$

$$0.055x + 56 - 56 = 276 - 56$$

$$0.055x = 220$$

$$\frac{0.055^1 x}{0.055^1} = \frac{220}{0.055}$$

$x = \$4000$ invested at 2%

$x + 1600 = 4000 + 1600 = \$5600$ invested at $3\frac{1}{2}\%$

5.   Let $x =$ the amount of money invested at 3% and $4x =$ the amount of money invested at $4\frac{1}{2}\%$ and $x + 900 =$ the amount of money invested at 6%.

$$3\%x + 4\frac{1}{2}\%(4x) + 6\%(x + 900) = 297$$

$$0.03x + 0.18x + 0.06x + 54 = 297$$

$$0.27x + 54 = 297$$

$$0.27x + 54 - 54 = 297 - 54$$

$$0.27x = 243$$

$$\frac{0.27^1 x}{0.27} = \frac{243}{0.27}$$

$$x = \$900 \text{ invested at } 3\%$$

$$4x = 4(900) = \$3600 \text{ invested at } 4\frac{1}{2}\%$$

$$x + 900 = 900 + 900 = \$1800 \text{ invested at } 6\%$$

6.   Let $x =$ the amount of money invested at $2\frac{3}{4}\%$ and $5x =$ the amount of money invested at $5\frac{1}{2}\%$.

$$2\frac{3}{4}\%x + 5\frac{1}{2}\%(5x) = \$181.50$$

$$0.0275x + 0.055(5x) = 181.50$$

$$0.0275x + 0.275x = 181.50$$

$$0.3025x = 181.50$$

$$\frac{0.3025^1 x}{0.3025^1} = \frac{181.50}{0.3025}$$

$$x = \$600 \text{ invested at } 2\frac{3}{4}\%$$

$$5x = 5(600) = \$3000 \text{ invested at } 5\frac{1}{2}\%$$

7. Let $x =$ the amount of money invested at 5% and $x - 1300 =$ the amount of money invested at 9%.

$$5\%x + 9\%(x - 1300) = 219$$

$$0.05x + 0.09(x - 1300) = 219$$

$$0.05x + 0.09x - 117 = 219$$

$$0.14x - 117 = 219$$

$$0.14x - 117 + 117 = 219 + 117$$

$$0.14x = 336$$

$$\frac{\cancel{0.14}^{1}x}{\cancel{0.14}^{1}} = \frac{336}{0.14}$$

$$x = \$2400 \text{ invested at } 5\%$$

$$x - 1300 = 2400 - 1300 = \$1100 \text{ invested at } 9\%$$

8. Let $x =$ the amount of money invested at 7% and $12000 - x =$ the amount of money invested at 5%.

$$7\%x = 5\%(12,000 - x)$$

$$0.07x = 0.05(12,000 - x)$$

$$0.07x = 600 - 0.05x$$

$$0.07x + 0.05x = 600 - 0.05x + 0.05x$$

$$0.12x = 600$$

$$\frac{\cancel{0.12}^{1}x}{\cancel{0.12}^{1}} = \frac{600}{0.12}$$

$$x = \$5000 \text{ invested at } 7\%$$

$$12{,}000 - x = 12{,}000 - 5000 = 7000 \text{ invested at } 5\%$$

9. Let $x =$ the amount of money invested at 1% and $3x =$ the amount of money invested at 4% and $x + 400 =$ the amount of money invested at 3%.

$$1\%x + 4\%(3x) + 3\%(x + 400) = \$108$$

$$0.01x + 0.04(3x) + 0.03(x + 400) = 108$$

$$0.01x + 0.12x + 0.03x + 12 = 108$$

$$0.16x + 12 = 108 - 12$$

$$0.16x = 96$$

$$\frac{\cancel{0.16}^{1}x}{\cancel{0.16}^{1}} = \frac{96}{0.16}$$

$$x = \$600 \text{ invested at } 1\%$$

$$3x = 3(600) = \$1800 \text{ invested at } 4\%$$

$$x + 400 = 600 + 400 = \$1000 \text{ invested at } 3\%$$

10. Let $x =$ the amount of money to be invested at 4%.

$$6\%(\$3000) + 4\%x = \$372$$

$$0.06(3000) + 0.04x = 372$$

$$180 + 0.04x = 372$$

$$180 - 180 + 0.04x = 372 - 180$$

$$0.04x = 192$$

$$\frac{\cancel{0.04}^{1}x}{\cancel{0.04}^{1}} = \frac{192}{0.04}$$

$$x = \$4800$$

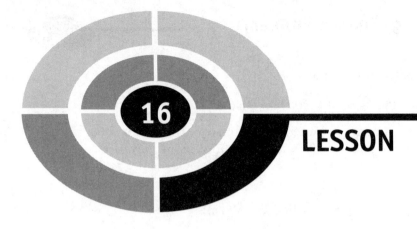

Solving Lever Problems

One of the oldest machines known to humans is the lever. The principles of the lever are studied in physics. Most people are familiar with the simplest kind of lever, known as the seesaw or teeterboard, often seen in parks.

The lever is a board placed on a **fulcrum** or point of support. On a seesaw, the fulcrum is in the center of the board. A child sits at either end of the board. If one child is heavier than the other child, he or she can sit closer to the center in order to balance the seesaw. This is the basic principle of the lever.

In general, the weights are placed on the ends of the board, and the distance the weight is from the fulcrum is called the length or arm. The basic principle of the lever is that the weight times the length of the arm on the left side of the lever is equal to the weight times the length of the arm on the right side of the lever, or $WL = wl$. See Figure 16-1.

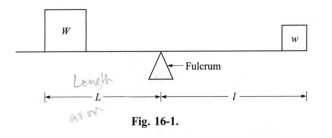

Fig. 16-1.

Given any of the three variables, you can set up an equation and solve for the fourth one. Unless otherwise specified, assume the fulcrum is in the center of the lever.

EXAMPLE: Bill weighs 120 pounds and sits on a seesaw 3 feet from the fulcrum. Where must Mary, who weighs 96 pounds, sit to balance it?

SOLUTION:

GOAL: You are being asked to find the distance from the fulcrum Mary needs to sit to balance the seesaw.

STRATEGY: Use the formula $WL = wl$ where $W = 120$, $L = 3$, $w = 96$, $l = x$.

$$WL = wl$$
$$120(3) = 96(x)$$

See Figure 16-2.

IMPLEMENTATION: Solve the equation:

$$120(3) = 96(x)$$
$$360 = 96x$$

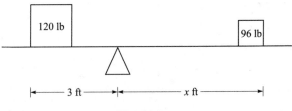

Fig. 16-2.

$$\frac{360}{96} = \frac{\cancel{96}^{1} x}{\cancel{96}^{1}}$$

$$3.75 = x$$

Hence she must sit 3.75 feet from the fulcrum.

EVALUATION: Check the equation:

$$WL = wl$$

$$120(3) = 96(3.75)$$

$$360 = 360$$

The fulcrum of a lever does not have to be at its center.

EXAMPLE: The fulcrum of a lever is 3 feet from the end of a 10-foot lever. On the short end rests an 84-pound weight. How much weight must be placed on the other end to balance the lever?

SOLUTION:

GOAL: You are being asked to find how much weight is needed to balance the lever.

STRATEGY: Let $x =$ the weight of the object needed.

$$WL = wl$$

$$84(3) = x(7)$$

See Figure 16-3.

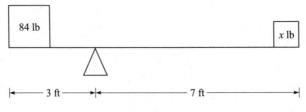

Fig. 16-3.

IMPLEMENTATION: Solve the equation:

$$84(3) = 7(x)$$
$$252 = 7x$$
$$\frac{252}{7} = \frac{\cancel{7}^1 x}{\cancel{7}^1}$$
$$36 = x$$

36 pounds needs to be placed at the 7-foot end to balance the lever.

EVALUATION:

$$WL = wl$$
$$84(3) = 36(7)$$
$$252 = 252$$

EXAMPLE: Where should the fulcrum be placed on an 18-foot lever with a 36-pound weight on one end and a 64-pound weight on the other end?

SOLUTION:

GOAL: You are being asked to find the placement of the fulcrum so that the lever is balanced.

STRATEGY: Let $x =$ the length of the lever from the fulcrum to the 36-pound weight and $(18 - x) =$ the length of the lever from the fulcrum to the 64-pound weight. See Figure 16-4.

The equation is

$$WL = wl$$
$$36x = 64(18 - x)$$

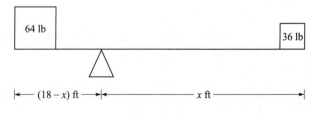

Fig. 16-4.

STRATEGY: Solve the equation:

$$36x = 64(18 - x)$$

$$36x = 1152 - 64x$$

$$36x + 64x = 1152 - 64x + 64x$$

$$100x = 1152$$

$$\frac{\cancel{100}^1 x}{\cancel{100}^1} = \frac{1152}{100}$$

$$x = 11.52$$

Hence the fulcrum must be placed 11.52 feet from the 36-pound weight.

EVALUATION: Check the equation:

$$WL = wl$$

$$36(11.52) = 64(18 - 11.52)$$

$$414.72 = 414.72$$

    You can place 3 or more weights on a lever and it still can be balanced. If 4 weights are used, two on each side, the equation is

$$W_1L_1 + W_2L_2 = w_1l_1 + w_2l_2$$

EXAMPLE: On a 16-foot seesaw Fred, weighing 80 pounds, sits on one end. Next to Fred sits Bill, weighing 84 pounds. Bill is 4 feet from the fulcrum. On the other side at the end sits Pete, weighing 95 pounds. Where should Sam, weighing 75 pounds, sit in order to balance the seesaw?

SOLUTION:

GOAL: You are being asked to find the distance from the fulcrum where Sam should sit in order to balance the seesaw.

STRATEGY: Let $x =$ the distance from the fulcrum where Sam needs to sit. See Figure 16-5.

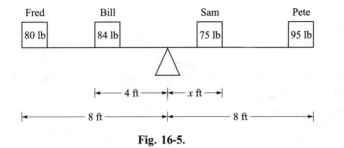

Fig. 16-5.

The equation is

$$W_1L_1 + W_2L_2 = w_1l_1 + w_2l_2$$

$$80(8) + 84(4) = 95(8) + 75(x)$$

$$640 + 336 = 760 + 75x$$

$$976 = 760 + 75x$$

$$976 - 760 = 760 - 760 + 75x$$

$$216 = 75x$$

$$\frac{216}{75} = \frac{\cancel{75}^1 x}{\cancel{75}}$$

$$2.88 = x$$

Sam needs to sit 2.88 feet from the fulcrum.

EVALUATION: Check the equation:

$$W_1L_1 + W_2L_2 = w_1l_1 + w_2l_2$$

$$80(8) + 84(4) = 95(8) + 75(2.88)$$

$$640 + 336 = 760 + 216$$

$$976 = 976$$

Try These

1. Brooke weighs 84 pounds and sits 6 feet from the fulcrum of a seesaw. If Mary weighs 72 pounds, how far should she sit from the fulcrum to balance the seesaw?

2. At one end of a lever is a 12-pound weight which is 8 inches from the fulcrum. How much weight should be placed on the other end 10 inches from the fulcrum to balance the lever?

3. A person places a lever under an 80-pound rock that is 3 feet from the fulcrum. How much pressure in pounds must the person place on the other end of the lever if it is 5 feet from the fulcrum to lift the rock?

4. Where should the fulcrum be placed under a 6-foot lever if there is a 24-pound weight on one end and a 36-pound weight on the other end in order to balance the lever?

5. On a 24-foot seesaw Beth, weighing 72 pounds, sits on one end. Mary, weighing 90 pounds, sits in front of her 8 feet from the fulcrum. On the other side at the end sits Megan, who weighs 80 pounds. Where should Tina, who weighs 96 pounds, sit in order to balance the seesaw?

SOLUTIONS:

1. Let $x =$ the distance Mary should sit from the fulcrum.

$$WL = wl$$

$$84(6) = (72)x$$

$$504 = 72x$$

$$\frac{504}{72} = \frac{\cancel{72}^{1}x}{\cancel{72}}$$

$$7 = x$$

Mary should sit 7 feet from the fulcrum.

2. Let $x =$ the weight placed on the other side of the lever.

$$WL = wl$$

$$12(8) = x(10)$$

$$96 = 10x$$

$$\frac{96}{10} = \frac{\cancel{10}^{1}x}{\cancel{10}^{1}}$$

$$9.6 = x$$

A weight of 9.6 pounds should be placed 10 inches from the fulcrum to balance the lever.

3. Let $x =$ the pressure in pounds needed to move the rock.

$$WL = wl$$

$$80(3) = x(5)$$

$$240 = 5x$$

$$\frac{240}{5} = \frac{\cancel{5}^{1}x}{\cancel{5}^{1}}$$

$$48 = x$$

48 pounds of pressure is needed to move the rock.

4. Let $x =$ the distance from the fulcrum where a 24-pound weight sits and $6 - x =$ the distance from the fulcrum the 36-pound weight sits.

$$WL = wl$$

$$24(x) = 36(6 - x)$$

$$24x = 216 - 36x$$

$$24x + 36x = 216 - 36x + 36x$$

$$60x = 216$$

$$\frac{\cancel{60}^{1}x}{\cancel{60}^{1}} = \frac{216}{60}$$

$$x = 3.6 \text{ feet}$$

The fulcrum should be placed 3.6 feet from the end of the lever which has the 24-pound weight.

5. Let $x =$ the distance from the fulcrum where Tina should sit.

$$W_1 L_1 + W_2 L_2 = w_1 l_1 + w_2 l_2$$

$$72(12) + 90(8) = 96x + 80(12)$$

$$864 + 720 = 96x + 960$$

$$1584 = 96x + 960$$

$$1584 - 960 = 96x + 960 - 960$$

$$624 = 96x$$

$$\frac{624}{96} = \frac{96^1 x}{96^1}$$

$$6.5 = x$$

Tina should sit 6.5 feet from the fulcrum.

Solving Work Problems

Work problems involve people doing a job. For example, if Frank can cut a lawn in 2 hours and his younger brother can cut the same lawn in 3 hours, how long will it take them to cut the grass if they both work together? In this case, we have two people doing the same job at the same time but at different rates.

Another type of problem involves pipes filling or draining bodies of water such as tanks, reservoirs, or swimming pools, at different rates. For example, if one pipe can fill a large tank in 5 hours, and a smaller pipe can fill the tank in 3 hours, how long would it take to fill the tank if both pipes are turned on at the same time? Again, we have two pipes doing the same job at different rates.

The basic principle is that the amount of work done by one person, machine, or pipe, plus the amount of work done by the second person, machine, or pipe, is equal to the total amount of work done in a given specific time. Also, the amount of work done by a single person, machine, or pipe is equal to the rate times the time, that is,

Rate × Time = Amount of work done

$Rate = \dfrac{Work}{time}$

EXAMPLE: Frank can cut a lawn in 2 hours. His brother Jeff can cut the same lawn in 3 hours. How long will it take them if they cut the lawn at the same time?

SOLUTION:

GOAL: You are being asked to find the time in hours it will take both boys to cut the grass if they work together.

STRATEGY: Let $x =$ the time it takes them if they work together. Now, in one hour, Frank can cut $\frac{1}{2}$ of the lawn and Jeff can cut $\frac{1}{3}$ of the lawn.

	Rate	× Time	= Amount of work done
Frank	$\dfrac{1}{2}$	x	$\dfrac{1}{2}x$
Jeff	$\dfrac{1}{3}$	x	$\dfrac{1}{3}x$

Frank does $\frac{1}{2}x$ or $\frac{x}{2}$ amount of work and Jeff does $\frac{1}{3}x$ or $\frac{x}{3}$ amount of work. These are the fractional parts of work done by each. Then the total amount of work done is 100% or 1. The equation is

$$\frac{x}{2} + \frac{x}{3} = 1$$

The LCD of 2 and 3 is 6, so clear fractions:

$$\frac{6^{3}}{1} \cdot \frac{x}{2^{1}} + \frac{6^{2}}{1} \cdot \frac{x}{3^{1}} = 6 \cdot 1$$

$$3x + 2x = 6$$

$$5x = 6$$

$$\frac{5^{1}x}{5} = \frac{6}{5}$$

$$x = 1.2 \text{ hours}$$

Hence, if both work together, then they can cut the lawn in 1.2 hours.

EVALUATION: Check the equation:

$$x = 1.2 \qquad \frac{1.2}{2} + \frac{1.2}{3} = 1$$
$$0.6 + 0.4 = 1$$
$$1 = 1$$

EXAMPLE: One pipe can fill a large tank in 5 hours and a smaller pipe can fill the same tank in 3 hours. How long will it take both pipes to fill the tank if they are turned on at the same time?

SOLUTION:

GOAL: You are being asked to find the time in hours it would take to fill the tank if both pipes are filling the tank at the same time.

STRATEGY: Let $x =$ the time it takes to fill the tank with both pipes. In one hour, the first pipe does $\frac{1}{5}$ of the work and the second pipe does $\frac{1}{3}$ of the work.

	Rate	×	Time	=	Amount of work done
First pipe	$\frac{1}{5}$		x		$\frac{1}{5}x$
Second pipe	$\frac{1}{3}$		x		$\frac{1}{3}x$

Again, the total amount of work done is 100% or 1.

The equation is

$$\frac{x}{5} + \frac{x}{3} = 1$$

IMPLEMENTATION: Solve the equation:

$$\frac{x}{5} + \frac{x}{3} = 1$$

LESSON 17   Solving Work Problems

The LCD is 15.

$$\frac{\cancel{15}^3}{1} \cdot \frac{x}{\cancel{5}^1} + \frac{\cancel{15}^5}{1} \cdot \frac{x}{\cancel{3}^1} = 15 \cdot 1$$

$$3x + 5x = 15$$

$$8x = 15$$

$$x = \frac{15}{8} \text{ or } 1.875 \text{ hours}$$

Hence, if both pipes are turned on at the same time, it would take 1.875 hours.

EVALUATION: Check the equation:

$$\frac{x}{5} + \frac{x}{3} = 1$$

$$\frac{1.875}{5} + \frac{1.875}{3} = 1$$

$$0.375 + 0.625 = 1$$

$$1 = 1$$

As you can see, both types of problems can be done using the same strategy. The next examples show some variations of work problems.

EXAMPLE: A person can clean a small office building in 8 hours and her assistant can clean the same building in 12 hours. If, on a certain day, the assistant shows up two hours late and starts to work, how long will it take both people to clean the building?

SOLUTION:

GOAL: You are being asked to find the time it takes both workers to clean the building.

STRATEGY: Let $x =$ the time it takes to clean the rest of the building when both people are working.

	Rate	×	Time	= Amount of work done
First cleaner	$\dfrac{1}{8}$		x	$\dfrac{1}{8}x$
Assistant cleaner	$\dfrac{1}{12}$		x	$\dfrac{1}{12}x$

Since the assistant starts two hours later, the first cleaner has already done $2 \times \frac{1}{8}$ or $\frac{2}{8}$ of the work. Hence, the equation is

$$\frac{2}{8} + \frac{1}{8}x + \frac{1}{12}x = 1$$

IMPLEMENTATION: Solve the equation:

$$\frac{2}{8} + \frac{1}{8}x + \frac{1}{12}x = 1$$

The LCD is 24.

$$\frac{24^3}{1} \cdot \frac{2}{8^1} + \frac{24^3}{1} \cdot \frac{x}{8^1} + \frac{24^2}{1} \cdot \frac{x}{12^1} = 24 \cdot 1$$

$$6 + 3x + 2x = 24$$

$$6 + 5x = 24$$

$$6 - 6 + 5x = 24 - 6$$

$$\frac{5^1 x}{5^1} = \frac{18}{5}$$

$$x = 3.6 \text{ hours}$$

Since the person has already worked two hours, the time it takes to clean the whole building is $2 + 3.6 = 5.6$ hours.

EVALUATION: Check the equation:

$$\frac{2}{8} + \frac{1}{8}x + \frac{1}{12}x = 1$$

Use $x = 3.6$.

$$\frac{2}{8} + \frac{3.6}{8} + \frac{3.6}{12} = 1$$

$$0.25 + 0.45 + 0.3 = 1$$

$$1 = 1$$

EXAMPLE: A swimming pool can be filled in 6 hours and drained in 15 hours. How long will it take to fill the pool if the owner has forgotten to close the drain valve?

SOLUTION:

GOAL: You are being asked how long in hours it will take to fill the pool if the drain is left open.

STRATEGY: Let $x =$ the time in hours it takes to fill the pool.

	Rate	×	Time	= Amount of work done
Fill pool	$\frac{1}{6}$		x	$\frac{1}{6}x$
Empty pool	$\frac{1}{15}$		x	$\frac{1}{15}x$

Since the drain is emptying the pool, the equation is

$$\frac{1}{6}x - \frac{1}{15}x = 1$$

IMPLEMENTATION: Solve the equation:

$$\frac{1}{6}x - \frac{1}{15}x = 1$$

The LCD is 30.

$$\frac{\cancel{30}^5}{1} \cdot \frac{x}{\cancel{6}^1} - \frac{\cancel{30}^2}{1} \cdot \frac{x}{\cancel{15}^1} = 30 \cdot 1$$

$$5x - 2x = 30$$

$$3x = 30$$

$$\frac{\cancel{3}^1 x}{\cancel{3}} = \frac{30}{3}$$

$$x = 10 \text{ hours}$$

Hence it will take 10 hours to fill the pool.

EVALUATION: Check the equation:

$$\frac{x}{6} - \frac{x}{15} = 1$$

$$x = 10 \qquad \frac{10}{6} - \frac{10}{15} = 1$$

$$1.67 - 0.67 = 1$$

$$1 = 1$$

EXAMPLE: Mary can do a job in 80 minutes. Working with Jane, both can do the job in 30 minutes. How long will it take Jane to do the job alone?

SOLUTION:

GOAL: You are being asked to find the time in minutes it takes for Jane to complete the job alone.

STRATEGY: Let $x =$ the time it takes Jane to complete the job.

	Rate $\times$	Time	$=$ Amount of work done
Mary	$\dfrac{1}{80}$	30	$\dfrac{30}{80}$
Jane	$\dfrac{1}{x}$	30	$\dfrac{30}{x}$

The equation is

$$\frac{30}{80} + \frac{30}{x} = 1$$

IMPLEMENTATION: Solve the equation:

$$\frac{30}{80} + \frac{30}{x} = 1$$

The LCD $= 80x$

$$\frac{\cancel{80}x}{1} \cdot \frac{30}{\cancel{80}} + 80 \cdot 30 = 80x \cdot 1$$

$$30x + 80 \cdot 30 = 80x$$

$$30x + 2400 = 80x$$

$$30x - 30x + 2400 = 80x - 30x$$

$$2400 = 50x$$

$$\frac{2400}{50} = \frac{\cancel{50}^1 x}{\cancel{50}^1}$$

$$48 = x$$

Hence, it will take Jane 48 minutes to do the job alone.

EVALUATION: Check the equation:

$$\frac{30}{80} + \frac{30}{48} = 1$$

$$\frac{30}{80} + \frac{30}{48} = 1$$

$$0.375 + 0.625 = 1$$

$$1 = 1$$

Try These

1. June can paint a shed in 9 hours and her father can paint it in 6 hours. How long will it take if they both paint it together?

2. Mike can plow a field in 12 hours and his brother Phil can plow it in 15 hours. How long will it take them to plow it if they use two plows and work together?

3. Sarah can complete a project in 90 minutes and her sister Betty can complete it in 120 minutes. If they both work on the project at the same time, how long will it take them to complete the project?

4. One pipe can empty a tank in 45 minutes while a second pipe can empty it in 60 minutes. If both pipes are opened at the same time, how long will it take to drain the tank?

5. One faucet can fill a large tub in 32 minutes while another faucet can fill the tub in 48 minutes. How long will it take to fill the tub if both faucets are opened at the same time?

6. A pipe can fill a tank in 120 minutes while the drain can drain it in 150 minutes. If the drain is left open and the fill pipe is turned on, how long will it take to fill the tub?

7. George can complete a project in 75 minutes, and if George and Walter both work on the project, they can complete it in 45 minutes. How long will it take Walter to complete the project by himself?

8. Pipe A can fill a tank in 60 minutes, Pipe B can fill it in 80 minutes, and Pipe C can fill it in 90 minutes. If all three pipes are opened at the same time, how long will it take to fill the tank?

9. Mary can make a costume twice as fast as Sam can. If they both work together, they can make it in 6 hours. How long will it take Mary to make the costume if she works alone?

10. A man can seed a large field in 8 hours. His daughter can do the job in 6 hours. If she starts an hour after her father, how long will it take to seed the field?

SOLUTIONS:

1. Let $x=$ the time it takes if both people work together.

$$\frac{1}{9}x+\frac{1}{6}x=1 \qquad LCD=18$$

$$\frac{\cancel{18}^2}{1}\cdot\frac{x}{\cancel{9}^1}+\frac{\cancel{18}^3}{1}\cdot\frac{x}{\cancel{6}^1}=18\cdot1$$

$$2x+3x=18$$

$$5x=18$$

$$\frac{\cancel{5}^1 x}{\cancel{5}} = \frac{18}{5}$$

$$x = 3.6 \text{ hours}$$

2. Let $x =$ the time it takes if both people work together.

$$\frac{1}{12}x + \frac{1}{15}x = 1 \qquad \text{LCD} = 60$$

$$\frac{\cancel{60}^5}{1} \cdot \frac{x}{\cancel{12}^1} + \frac{\cancel{60}^4}{1} \cdot \frac{x}{\cancel{15}^1} = 60 \cdot 1$$

$$5x + 4x = 60$$

$$9x = 60$$

$$\frac{\cancel{9}x}{\cancel{9}} = \frac{60}{9}$$

$$x = 6\frac{2}{3} \text{ or } 6.67 \text{ hours}$$

3. Let $x =$ the time it takes if both people work together.

$$\frac{1}{90}x + \frac{1}{120}x = 1 \qquad \text{LCD} = 360$$

$$\frac{360^4}{1} \cdot \frac{x}{\cancel{90}^1} + \frac{360^3}{1} \cdot \frac{x}{\cancel{120}^1} = 360 \cdot 1$$

$$4x + 3x = 360$$

$$7x = 360$$

$$\frac{\cancel{7}^1 x}{\cancel{7}} = \frac{360}{7}$$

$$x = 51\frac{3}{7} \text{ or } 51.43 \text{ hours}$$

4. Let $x =$ the time it will take to fill the tank if both pipes are open at the same time.

$$\frac{1}{45}x + \frac{1}{60}x = 1 \qquad LCD = 180$$

$$\frac{180^4}{1} \cdot \frac{x}{45^1} + \frac{180^3}{1} \cdot \frac{x}{60^1} = 180 \cdot 1$$

$$4x + 3x = 180$$

$$7x = 180$$

$$\frac{7^1 x}{7^1} = \frac{180}{7}$$

$$x = 25\frac{5}{7} \text{ or } 25.71 \text{ minutes}$$

5. Let $x =$ the time it will take to fill the tub if both faucets are open.

$$\frac{1}{32}x + \frac{1}{48}x = 1 \qquad LCD = 96$$

$$\frac{96^3}{1} \cdot \frac{x}{32^1} + \frac{96^2}{1} \cdot \frac{x}{48^1} = 96 \cdot 1$$

$$3x + 2x = 96$$

$$5x = 96$$

$$\frac{5^1 x}{5^1} = \frac{96}{5}$$

$$x = 19.2 \text{ minutes}$$

6. Let $x =$ the time it will take to fill the tank.

$$\frac{1}{120}x - \frac{1}{150}x = 1 \qquad LCD = 600$$

$$\frac{600^5}{1} \cdot \frac{x}{120^1} - \frac{600^4}{1} \cdot \frac{x}{150^1} = 600 \cdot 1$$

$$5x - 4x = 600$$

$$x = 600 \text{ minutes}$$

7. Let $x =$ the time it will take Walter to complete the project.

$$\frac{45}{75} + \frac{45}{x} = 1 \qquad LCD = 75x$$

$$\cancel{75}^{1}x \cdot \frac{45}{\cancel{75}^{1}} + 75\cancel{x} \cdot \frac{45}{\cancel{x}} = 75x \cdot 1$$

$$45x + 3375 = 75x$$

$$45x - 45x + 3375 = 75x - 45x$$

$$3375 = 30x$$

$$\frac{3375}{30} = \frac{\cancel{30}^{1}x}{\cancel{30}^{1}}$$

$$112.5 \text{ minutes} = x$$

8. Let $x =$ the time it takes to fill the tank if all three pipes are open.

$$\frac{1}{60}x + \frac{1}{80}x + \frac{1}{90}x = 1 \qquad LCD = 720$$

$$\frac{\cancel{720}^{12}}{1} \cdot \frac{x}{\cancel{60}^{1}} + \frac{\cancel{720}^{9}}{1} \cdot \frac{x}{\cancel{80}^{1}} + \frac{\cancel{720}^{8}}{1} \cdot \frac{x}{\cancel{90}^{1}} = 720 \cdot 1$$

$$12x + 9x + 8x = 720$$

$$29x = 720$$

$$\frac{\cancel{29}^{1}x}{\cancel{29}^{1}} = \frac{720}{29}$$

$$x = 24.83 \text{ minutes}$$

9. Let $x =$ the time it takes Mary to make the costume and $2x =$ the time it takes Sam to make the costume.

$$\frac{6}{x} + \frac{6}{2x} = 1 \qquad LCD = 2x$$

$$2\cancel{x} \cdot \frac{6}{\cancel{x}} + 2\cancel{x} \cdot \frac{6}{2\cancel{x}} = 2x \cdot 1$$

$$12 + 6 = 2x$$

$$18 = 2x$$

$$\frac{18}{2} = \frac{2^1 x}{2^1}$$

$$9 \text{ hours} = x$$

10. Let $x =$ the time it takes both people to seed the field.

$$\frac{1}{8} + \frac{1}{8}x + \frac{1}{6}x = 1 \qquad LCD = 24$$

$$\frac{24^3}{1} \cdot \frac{1}{8^1} + \frac{24^3}{1} \cdot \frac{x}{8^1} + \frac{24^4}{1} \cdot \frac{x}{6^1} = 24 \cdot 1$$

$$3 + 3x + 4x = 24$$

$$3 + 7x = 24$$

$$3 - 3 + 7x = 24 - 3$$

$$7x = 21$$

$$x = 3 \text{ hours}$$

Since the man has already worked an hour before his daughter started, it will take $1 + 3 = 4$ hours to complete the job.

Quiz 4

1. Juan is 5 years older than his sister. Ten years ago, he was twice as old as his sister. Find his present age.
   - (a)   15
   - (b)   5
   - (c)   20
   - (d)   10

2. Two people leave an office and drive in opposite directions. If one person travels 5 miles per hour faster than the other and at the end of 3 hours they are 207 miles apart, find the speed of the slower driver.
   - (a)   32 miles per hour
   - (b)   35 miles per hour
   - (c)   37 miles per hour
   - (d)   42 miles per hour

3. A boat travels to a destination upstream in 2 hours and returns in 1.5 hours. If the current is 5 miles per hour, find the speed of the boat on its speedometer.
   (a) 30 miles per hour
   (b) 45 miles per hour
   (c) 40 miles per hour
   (d) 35 miles per hour

4. A candy store owner wants to mix some candy costing $1.25 a pound with some candy costing $1.45 a pound to make 50 pounds of candy costing $1.30 a pound. How much of the $1.25 a pound candy should she use?
   (a) 35 pounds
   (b) 37.5 pounds
   (c) 15 pounds
   (d) 12.5 pounds

5. A craftsperson wishes to make an alloy from an alloy which is 40% gold and another which is 60% gold. How many ounces of the 40% alloy need to be used to get 5 ounces that are 55% gold?
   (a) 4.75 ounces
   (b) 3.25 ounces
   (c) 1.25 ounces
   (d) 2.75 ounces

6. An investor has part of $1200 invested at 8% and the rest invested at 5%. If the total yearly interest on the money is $87.60, how much does the person have invested at 5%?
   (a) $280
   (b) $440
   (c) $920
   (d) $760

7. A person has a certain sum of money invested at 3% and twice that amount invested at 7%. If the total interest from the investment is $119, find the amount of money invested at 7%.
   (a) $1750
   (b) $3500
   (c) $7000
   (d) $1400

8. Bernie, who weighs 95 pounds, sits on one end of a 20-foot seesaw. How far from the fulcrum must Billy, who weighs 118.75 pounds, sit to balance the seesaw?
   (a)   9.5 feet
   (b)   8 feet
   (c)   6.5 feet
   (d)   7 feet

9. A small pipe can fill a tank in 20 minutes whereas a large pipe can fill it in 12 minutes. If both pipes are opened at the same time, how long will it take to fill the tank?
   (a)   6 minutes
   (b)   7.5 minutes
   (c)   8 minutes
   (d)   8.5 minutes

10. Joan can paint a large auditorium ceiling in 6 hours. Her sister can paint the ceiling in 4 hours. If they both work together, how long will it take them to paint the ceiling?
   (a)   4.2 hours
   (b)   3.8 hours
   (c)   1.6 hours
   (d)   2.4 hours

Systems of Equations

Two equations with two variables, usually x and y, are called a **system** of equations. For example,

$$x - y = 3$$
$$2x + y = 12$$

is called a system of equations. The solution to a system of equations consists of the values for the two variables which, when substituted in the equations, make both equations true at the same time. In this case, the solution for the system shown is $x = 5$ and $y = 2$. This can be shown as follows:

$$x - y = 3 \qquad\qquad 2x + y = 12$$
$$5 - 2 = 3 \qquad\qquad 2(5) + 2 = 12$$
$$3 = 3 \qquad\qquad 12 = 12$$

In other words, in order to solve a system of equations, it is necessary to find a value for x and a value for y which, when substituted in the equations, makes them both true.

There are several ways to solve a system of equations. The method used here is called the **substitution** method. You can use these steps:

Step 1: Select one equation and solve it for one variable in terms of the other variable.

Step 2: Substitute this expression for the variable in the **other** equation and solve it for the remaining variable.

Step 3: Select one of the equations, substitute the value for the variable found in Step 2, and solve for the other variable.

EXAMPLE: Solve the system

$$x - y = 3$$
$$2x + y = 12$$

SOLUTION:

Step 1: Select the first equation and solve it for x in terms of y.

$$x - y = 3$$
$$x - y + y = 3 + y$$
$$x = 3 + y$$

Step 2: Substitute $3 + y$ for x in the second equation and solve for y.

$$2x + y = 12$$
$$2(3 + y) + y = 12$$
$$6 + 2y + y = 12$$
$$6 + 3y = 12$$
$$6 - 6 + 3y = 12 - 6$$
$$3y = 6$$
$$\frac{\cancel{3}^1 y}{\cancel{3}^1} = \frac{6}{3}$$
$$y = 2$$

Step 3: Select $x - y = 3$, substitute $y = 2$ and solve for x.

$$x - y = 3$$
$$x - 2 = 3$$
$$x - 2 + 2 = 3 + 2$$
$$x = 5$$

Hence, the solution to the system is $x = 5$ and $y = 2$.

You can check the solution by substituting $x = 5$ and $y = 2$ in the other equation and seeing if it is true.

$$2x + y = 12$$
$$2(5) + 2 = 12$$
$$10 + 2 = 12$$
$$12 = 12$$

EXAMPLE: Solve the system

$$x + 3y = -7$$
$$3x - y = 9$$

SOLUTION:

Step 1: Solve the first equation for x.

$$x + 3y = -7$$
$$x + 3y - 3y = -7 - 3y$$
$$x = -7 - 3y$$

Step 2: Substitute $-7 - 3y$ in the second equation and solve for y.

$$3x - y = 9$$
$$3(-7 - 3y) - y = 9$$
$$-21 - 9y - y = 9$$
$$-21 - 10y = 9$$
$$-21 + 21 - 10y = 9 + 21$$

$$-10y = 30$$

$$\frac{-\cancel{10}^1 y}{-\cancel{10}^1} = \frac{30}{-10}$$

$$y = -3$$

Step 3:   Substitute -3 for y in the first equation and find the value for x.

$$x + 3y = -7$$

$$x + 3(-3) = -7$$

$$x - 9 = -7$$

$$x - 9 + 9 = -7 + 9$$

$$x = 2$$

You can check the solution by using $3x - y = 9$ when $x = 2$ and $y = -3$.

$$3x - y = 9$$

$$3(2) - (-3) = 9$$

$$6 + 3 = 9$$

$$9 = 9$$

When selecting an equation and a variable to solve for in Step 1, you should look for an equation which has a variable whose numerical coefficient is one. Since this is not always possible, you still can use the substitution method to solve the equation as shown in the next example.

EXAMPLE: Solve the system

$$5x - 7y = -31$$
$$4x + 3y = 1$$

SOLUTION:

Step 1:   Select the second equation and solve for y.

$$4x + 3y = 1$$

$$4x - 4x + 3y = 1 - 4x$$

$$3y = 1 - 4x$$

$$\frac{\cancel{3}^1 y}{\cancel{3}^1} = \frac{1 - 4x}{3}$$

$$y = \frac{1 - 4x}{3}$$

Step 2:   Substitute in the first equation.

$$5x - 7y = -31$$

$$5x - 7 \cdot \frac{(1 - 4x)}{3} = -31$$

Clear fractions.

$$3 \cdot 5x - \frac{\cancel{3}^1}{1} \cdot \frac{7(1 - 4x)}{\cancel{3}^1} = -31 \cdot 3$$

$$15x - 7(1 - 4x) = -93$$

$$15x - 7 + 28x = -93$$

$$43x - 7 + 7 = -93 + 7$$

$$43x = -86$$

$$\frac{\cancel{43}^1 x}{\cancel{43}^1} = \frac{-86}{43}$$

$$x = -2$$

Step 3:   Find y.

$$4x + 3y = 1$$

$$4(-2) + 3y = 1$$

$$-8 + 3y = 1$$

$$-8 + 8 + 3y = 1 + 8$$

$$3y = 9$$

$$\frac{\cancel{3}^1 y}{\cancel{3}^1} = \frac{9}{3}$$

$$y = 3$$

The solution is $x = -2$ and $y = 3$. You can check the answer.

Try These

Solve each system.

1. $5x + y = 7$
  $2x - 3y = -4$

2. $-4x + 5y = -2$
  $x + 3y = 9$

3. $4x - 3y = 1$
  $2x + y = 13$

4. $x = 2y$
  $3x + 4y = 40$

5. $2x - y = -11$
  $y = -x + 2$

SOLUTIONS:

1. Solve the first equation for y.

$$5x + y = 7$$
$$5x - 5x + y = 7 - 5x$$
$$y = 7 - 5x$$

Substitute in the second equation and solve for x.

$$2x - 3y = -4$$
$$2x - 3(7 - 5x) = -4$$
$$2x - 21 + 15x = -4$$
$$17x - 21 = -4$$
$$17x - 21 + 21 = -4 + 21$$
$$17x = 17$$
$$\frac{\cancel{17}^{1} x}{\cancel{17}^{1}} = \frac{17}{17}$$
$$x = 1$$

Find y:

$$5x + y = 7$$
$$5(1) + y = 7$$
$$5 + y = 7$$
$$5 - 5 + y = 7 - 5$$
$$y = 2$$

2. Solve the second equation for x.

$$x + 3y = 9$$
$$x + 3y - 3y = 9 - 3y$$
$$x = 9 - 3y$$

Substitute in the first equation and solve for y.

$$-4x + 5y = -2$$
$$-4(9 - 3y) + 5y = -2$$

$$-36 + 12y + 5y = -2$$

$$-36 + 17y = -2$$

$$-36 + 36 + 17y = -2 + 36$$

$$17y = 34$$

$$\frac{\cancel{17}^1 y}{\cancel{17}^1} = \frac{34}{17}$$

$$y = 2$$

Find x:

$$x + 3y = 9$$

$$x + 2(3) = 9$$

$$x + 6 = 9$$

$$x + 6 - 6 = 9 - 6$$

$$x = 3$$

3. Solve the second equation for y.

$$2x + y = 13$$

$$2x - 2x + y = 13 - 2x$$

$$y = 13 - 2x$$

Substitute in the first equation and solve for x.

$$4x - 3y = 1$$

$$4x - 3(13 - 2x) = 1$$

$$4x - 39 + 6x = 1$$

$$10x - 39 = 1$$

$$10x - 39 + 39 = 1 + 39$$

$$10x = 40$$

$$\frac{\cancel{10}^1 x}{\cancel{10}^1} = \frac{40}{10}$$

$$x = 4$$

Find y:

$$2x + y = 13$$

$$2(4) + y = 13$$

$$8 + y = 13$$

$$8 - 8 + y = 13 - 8$$

$$y = 5$$

4. Substitute for x in the second equation and find y since $x = 2y$.

$$3x + 4y = 40$$

$$3(2y) + 4y = 40$$

$$6y + 4y = 40$$

$$10y = 40$$

$$\frac{\cancel{10}^1 y}{\cancel{10}^1} = \frac{40}{10}$$

$$y = 4$$

Find x:

$$x = 2y$$

$$x = 2(4)$$

$$x = 8$$

5. Substitute for y in the first equation since $y = -x + 2$

$$2x - y = -11$$

$$2x - (-x + 2) = -11$$

$$2x + x - 2 = -11$$

$$3x - 2 = -11$$

$$3x - 2 + 2 = -11 + 2$$

$$3x = -9$$

$$\frac{\cancel{3}^1 x}{\cancel{3}^1} = \frac{-9}{3}$$

$$x = -3$$

Find y:

$$y = -x + 2$$
$$y = -(-3) + 2$$
$$y = 3 + 2$$
$$y = 5$$

Solving Word Problems Using Two Equations

If you need to review systems of equations, complete Refresher V.

Many of the previous types of problems can be solved using a system of two equations with two unknowns. The strategy used to solve problems using two equations is:

Step 1: Represent one of the unknowns as x and the other unknown as y.

Step 2: Translate the information about the variables into two equations using the two unknowns.

Step 3: Solve the system of equations for x and y.

In this lesson, a sample of each type of problem is solved by using a system of two equations with two unknowns. You will find the exact same problems in the previous sections. This was done so that you can compare the two methods (i.e., solving a problem using one equation versus solving a problem using two equations). For some types of problems, such as lever and work problems, it is better to use one equation.

EXAMPLE: One number is 8 more than another number and the sum of the two numbers is 26. Find the numbers.

SOLUTION:

GOAL: You are being asked to find two numbers.

STRATEGY: Let $x=$ the smaller number

and $y=$ the larger number

Since one number is 8 more than the other number, the first equation is

$$y = x + 8$$

Since the sum of the two numbers is 26, the second equation is

$$x + y = 26$$

IMPLEMENTATION: Solve the system

$$y = x + 8$$
$$x + y = 26$$

Substitute the value for y in the second equation and solve for x since $y = x + 8$.

$$x + y = 26$$
$$x + x + 8 = 26$$
$$2x + 8 = 26$$
$$2x + 8 - 8 = 26 - 8$$

$$2x = 18$$
$$\frac{2^1 x}{2^1} = \frac{18}{2}$$
$$x = 9$$

Find the other number.

$$y = x + 8$$
$$y = 9 + 8$$
$$y = 17$$

Hence, the numbers are 9 and 17.

EVALUATION: Check the second equation.

$$x + y = 26$$
$$9 + 17 = 26$$
$$26 = 26$$

EXAMPLE: The sum of the digits of a two-digit number is 15. If the digits are reversed, the new number is 9 more than the original number. Find the number.

SOLUTION:

GOAL: You are being asked to find a two-digit number.

STRATEGY: Let $x =$ the ten's digit

$y =$ the one's digit

$10x + y =$ original number

$10y + x =$ new number with digits reversed

Since the sum of the digits of the number is 15, the first equation is

$$x + y = 15$$

Since reversing the digits gives a new number which is 9 more than the original number, the equation is

$$(10x + y) + 9 = (10y + x)$$

IMPLEMENTATION: Solve the system

$$x + y = 15$$
$$10x + y + 9 = 10y + x$$

Solve the first equation for y, substitute in the second equation and find x.

$$x + y = 15$$
$$x - x + y = 15 - x$$
$$y = 15 - x$$
$$10x + y + 9 = 10y + x$$
$$10x + (15 - x) + 9 = 10(15 - x) + x$$
$$10x + 15 - x + 9 = 150 - 10x + x$$
$$9x + 24 = 150 - 9x$$
$$9x + 9x + 24 = 150 - 9x + 9x$$
$$18x + 24 - 24 = 150 - 24$$
$$18x = 126$$
$$\frac{\cancel{18}^{1}x}{\cancel{18}^{1}} = \frac{126}{18}$$
$$x = 7$$

Find y:

$$x + y = 15$$
$$7 + y = 15$$

$$7 - 7 + y = 15 - 7$$
$$y = 8$$

Hence, the number is 78.

EVALUATION: Check the information in the second equation.

Original number $= 78$

Reversed number $= 87$

Since 87 is 9 more than 78, the answer is correct.

EXAMPLE: A person has 8 coins consisting of quarters and dimes. If the total amount of this change is $1.25, how many of each kind of coin are there?

SOLUTION:

GOAL: You are being asked to find how many coins are quarters and how many coins are dimes.

STRATEGY: Let $x =$ the number of quarters

$y =$ the number of dimes

$25x =$ the value of the quarters

and $10y =$ the value of the dimes

Since there are 8 coins, the first equation is

$$x + y = 8$$

Since the total values of the quarters plus the dimes is $1.25 or 125¢, the second equation is

$$25x + 10y = 125$$

IMPLEMENTATION: Solve the system

$$x + y = 8$$
$$25x + 10y = 125$$

Find the value of y in the first equation. Substitute it in the second equation and solve for x.

$$x + y = 8$$

$$x - x + y = 8 - x$$

$$y = 8 - x$$

$$25x + 10y = 125$$

$$25x + 10(8 - x) = 125$$

$$25x + 80 - 10x = 125$$

$$15x + 80 = 125$$

$$15x + 80 - 80 = 125 - 80$$

$$15x = 45$$

$$\frac{\cancel{15}^{1} x}{\cancel{15}} = \frac{45}{15}$$

$$x = 3$$

Find y:

$$x + y = 8$$

$$3 + y = 8$$

$$3 - 3 + y = 8 - 3$$

$$y = 5$$

Hence, there are 3 quarters and 5 dimes.

EVALUATION: Find the values of each and see if their sum is $1.25.

$$3 \text{ quarters} = 3 \times \$0.25 = \$0.75$$

$$5 \text{ dimes} = 5 \times \$0.10 = \$0.50$$

$$\$0.75 + \$0.50 = \$1.25$$

EXAMPLE: Bill is 8 years older than his brother. In 3 years, Bill will be twice as old as his brother. Find their present ages.

SOLUTION:

GOAL: You are being asked to find the present ages of Bill and his brother.

STRATEGY: Let x = Bill's age

$$y = \text{his brother's age}$$

$$x + 3 = \text{Bill's age in 3 years}$$

$$y + 3 = \text{his brother's age in 3 years}$$

Since Bill is 8 years older than his brother, the first equation is

$$x = y + 8$$

In 3 years, Bill will be twice as old as his brother, so the second equation is

$$x + 3 = 2(y + 3)$$

IMPLEMENTATION: Solve the system

$$x = y + 8$$

$$x + 3 = 2(y + 3)$$

Substitute the value of x in the second equation and solve for y since $x = y + 8$.

$$x + 3 = 2(y + 3)$$

$$(y + 8) + 3 = 2(y + 3)$$

$$y + 8 + 3 = 2y + 6$$

$$y + 11 = 2y + 6$$

$$y - y + 11 = 2y - y + 6$$

$$11 = y + 6$$

$$11 - 6 = y + 6 - 6$$

$$5 = y$$

Select the first equation, let $y = 5$, and solve for x.

$$x = y + 8$$

$$x = 5 + 8$$

$$x = 13$$

Hence, Bill is 13 years old and his brother is 5 years old.

EVALUATION: Check the second equation for $x = 13$ and $y = 5$

$$x + 3 = 2(y + 3)$$

$$13 + 3 = 2(5 + 3)$$

$$16 = 2(8)$$

$$16 = 16$$

EXAMPLE: A person walked from his home to the football stadium at a rate of 3 miles per hour. After the game, he rode the bus back the same way at a rate of 22 miles per hour. If the total trip took 2 hours, how far did he walk?

SOLUTION:

GOAL: You are being asked to find the distance the person walked to the stadium.

STRATEGY: Let $x =$ the time the person walked

and $y =$ the time the person rode the bus

Since the total time is 2 hours, the first equation is

$$x + y = 2$$

Since the distances are equal and $D = RT$, the second equation is

$$3x = 22y$$

IMPLEMENTATION: Solve the system

$$x + y = 22$$
$$3x = 22y$$

Solve the first equation for y and substitute the value in the second equation, and then solve for x.

$$x + y = 2$$
$$x - x + y = 2 - x$$
$$y = 2 - x$$

Then

$$3x = 22y$$

$$3x = 22(2 - x)$$

$$3x = 44 - 22x$$

$$3x + 22x = 44 - 22x + 22x$$

$$25x = 44$$

$$\frac{25^1 x}{25} = \frac{44}{25}$$

$$x = 1.76 \text{ hours}$$

Find the distance using $D = RT$.

$$D = RT$$

$$D = 3(1.76)$$

$$= 5.28 \text{ miles}$$

EVALUATION: The time he rode the bus is $2 - 1.76 = 0.24$ hours. The distance is $D = RT$.

$$D = RT$$

$$D = 22(0.24)$$

$$= 5.28 \text{ miles}$$

EXAMPLE: A merchant mixes some coffee costing $4 a pound with some coffee costing $3 a pound. How much of each must be used in order to make 20 pounds of mixture costing $3.75 a pound?

SOLUTION:

GOAL: You are being asked to find how much of each coffee should be used.

STRATEGY: Let $x =$ the amount of $4 coffee used

and $y =$ the amount of $3 coffee used

Since the total amount of the mixture is 20 pounds, the first equation is

$$x + y = 20$$

Since the cost of the mixture is \$3.75, the second equation is

$$4x + 3y = 20(3.75)$$

IMPLEMENTATION: Solve the system

$$x + y = 20$$
$$4x + 3y = 20(3.75)$$

Solve the first equation for x. Substitute in the second equation and solve for y.

$$x + y = 20$$
$$x + y - y = 20 - y$$
$$x = 20 - y$$

Substitute:

$$4x + 3y = 20(3.75)$$
$$4(20 - y) + 3y = 20(3.75)$$
$$80 - 4y + 3y = 75$$
$$80 - y = 75$$
$$80 - 80 - y = 75 - 80$$
$$-y = -5$$
$$\frac{-y}{-1} = \frac{-5}{-1}$$
$$y = 5 \text{ pounds}$$

Solve for x:

$$x + y = 20$$

$$x + 5 = 20$$

$$x + 5 - 5 = 20 - 5$$

$$x = 15 \text{ pounds}$$

Hence, 15 pounds of the $4 coffee are needed and 5 pounds of the $3 coffee are needed.

EVALUATION: Check the second equation.

$$4x + 3y = 20(3.75)$$

$$4(15) + 3(5) = 75$$

$$60 + 15 = 75$$

$$75 = 75$$

EXAMPLE: A person has $5000 to invest and decides to invest part of it at 4% and the rest of it at $6\frac{1}{2}$%. If the total interest for the year is $245, how much does the person have invested at each rate?

SOLUTION:

GOAL: You are being asked to find the amounts of money invested at each rate.

STRATEGY: Let $x =$ the amount of money invested at 4%

and $y =$ the amount of money invested at $6\frac{1}{2}$%

Since the total amount of money is $5000, the first equation is

$$x + y = \$5000$$

Since the total interest is \$245, the second equation is

$$4\%x + 6\frac{1}{2}\%y = 245$$

IMPLEMENTATION: Solve the system

$$x + y = 5000$$
$$4\%x + 6\frac{1}{2}\%y = 245$$

Solve the first equation for x. Substitute in the second equation and solve for y.

$$x + y = 5000$$
$$x + y - y = 5000 - y$$
$$x = 5000 - y$$

Then

$$0.04x + 0.065y = 245$$
$$0.04(5000 - y) + 0.065y = 245$$
$$200 - 0.04y + 0.065y = 245$$
$$200 + 0.025y = 245$$
$$200 - 200 + 0.025y = 245 - 200$$
$$0.025y = 45$$
$$\frac{0.025^1 x}{0.025} = \frac{45}{0.025}$$
$$y = 1800$$

Find x:

$$x + y = 5000$$

$$x + 1800 = 5000$$

$$x + 1800 - 1800 = 5000 - 1800$$

$$x = \$3200$$

Hence, he has $3200 invested at 4% and $1800 invested at $6\frac{1}{2}$%.

EVALUATION: Check the second equation.

$$4\%x + 6\frac{1}{2}\%y = 245$$

$$0.04(3200) + 0.065(1800) = 245$$

$$128 + 117 = 245$$

$$245 = 245$$

Try These

Use two equations with two unknowns.

1. One number is 4 times another number. If their sum is 40, find the numbers.

2. The sum of the digits of a two-digit number is 14. If the digits are reversed, the new number is 18 more than the original number. Find the number.

3. A person has 18 coins, some of which are nickels and the rest of which are dimes. If the total amount of the coins is $1.30, find the number of nickels and dimes.

4. Matt is 4 times older than Mike. In 10 years, he will be twice as old as Mike. Find their ages.

5. A person made 40 small bags of candy, some of which sold for $0.50 and the rest sold for $0.75 each. If the total amount he made was $23.25, find the number of bags of each he sold.

SOLUTIONS:

1. Let $x =$ the first number

    and $y =$ the second number

    $$y = 4x$$
    $$x + y = 40$$

    Substitute for y in the second equation and solve for x.

    $$x + 4x = 40$$
    $$5x = 40$$
    $$\frac{\cancel{5}^{1} x}{\cancel{5}^{1}} = \frac{40}{5}$$
    $$x = 8$$

    Solve for y:

    $$y = 4x$$
    $$y = 4 \cdot 8$$
    $$y = 32$$

2. Let $x =$ the ten's digit

    and $y =$ the one's digit

    $$10x + y = \text{original number}$$
    $$10y + x = \text{number with digits reversed}$$
    $$x + y = 14$$
    $$10x + y + 18 = 10y + x$$

Solve for y in the first equation.

$$y = 14 - x$$

Substitute in the second equation and solve for x.

$$10x + (14 - x) + 18 = 10(14 - x) + x$$

$$10x + 14 - x + 18 = 140 - 10x + x$$

$$9x + 32 = 140 - 9x$$

$$9x + 9x + 32 = 140 - 9x + 9x$$

$$18x + 32 = 140$$

$$18x + 32 - 32 = 140 - 32$$

$$18x = 108$$

$$\frac{\cancel{18}^{1}x}{\cancel{18}} = \frac{108}{18}$$

$$x = 6$$

Solve for y:

$$x + y = 14$$

$$6 + y = 14$$

$$6 - 6 + y = 14 - 6$$

$$y = 8$$

3. Let $x =$ the number of nickels
    and $y =$ the number of dimes

$$x + y = 18$$

$$5x + 10y = 130$$

Solve for y in the first equation.

$$x - x + y = 18 - x$$
$$y = 18 - x$$

Substitute in the second equation and solve for x.

$$5x + 10(18 - x) = 130$$

$$5x + 180 - 10x = 130$$

$$-5x + 180 = 130$$

$$-5x + 180 - 180 = 130 - 180$$

$$-5x = -50$$

$$\frac{-\cancel{5}^{1}x}{-\cancel{5}} = \frac{-50}{-5}$$

$$x = 10 \text{ nickels}$$

Solve for y:

$$x + y = 18$$

$$10 + y = 18$$

$$10 - 10 + y = 18 - 10$$

$$y = 8 \text{ dimes}$$

4. Let $x =$ Matt's age
    and $y =$ Mike's age

$$x = 4y$$
$$x + 10 = 2(y + 10)$$

Substitute for x in the second equation and solve for y.

$$4y + 10 = 2y + 20$$

$$4y - 2y + 10 = 2y - 2y + 20$$

$$2y + 10 = 20$$

$$2y + 10 - 10 = 20 - 10$$

$$2y = 10$$

$$\frac{\cancel{2}^{1} y}{\cancel{2}^{1}} = \frac{10}{2}$$

$$y = 5$$

Solve for x:

$$x = 4y$$

$$x = 4 \cdot 5$$

$$x = 20$$

Matt is 20 and Mike is 5 years old.

5. Let $x =$ the number of bags of $0.50 candy

and $y =$ the number of bags of $0.75 candy

$$x + y = 40$$

$$50x + 75y = 2325$$

Solve for y in the first equation.

$$x - x + y = 40 - x$$

$$y = 40 - x$$

Substitute in the second equation and solve for x.

$$50x + 75(40 - x) = 2325$$

$$50x + 3000 - 75x = 2325$$

$$-25x + 3000 = 2325$$

$$-25x + 3000 - 3000 = 2325 - 3000$$

$$-25x = -675$$

$$\frac{-25^1 x}{-25^1} = \frac{-675}{-25}$$

$$x = 27$$

Find y:

$$x + y = 40$$

$$27 + y = 40$$

$$27 - 27 + y = 40 - 27$$

$$y = 13$$

Hence he sold 27 bags of $0.50 candy and 13 bags of $0.75 candy.

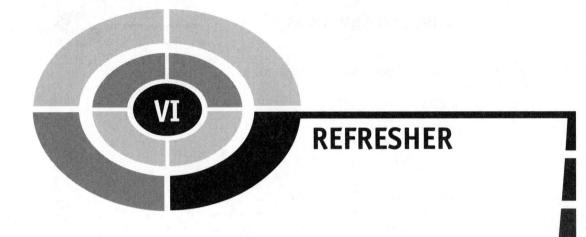

Quadratic Equations

An equation such as $2x^2 + 3x - 5 = 0$ is called a **quadratic equation** or a **second-degree equation**. There is one variable (usually x) and a second-degree term (usually x^2). There are several ways to solve quadratic equations. The method shown here will use **factoring**. *If you cannot factor trinomials, you will need to consult an algebra book to learn this skill.*

A quadratic equation can be written in standard form when the x^2 term is first, the x term is second, and the constant term is third. Also, zero is on the right side of the equation. For example, the quadratic equation $2x + x^2 = 8$ can be written in standard form as $x^2 + 2x - 8 = 0$. In order to solve a quadratic equation by factoring, you should follow these steps:

Step 1: Write the equation in standard form.
Step 2: Factor the left member of the equation.
Step 3: Set both factors equal to zero.
Step 4: Solve each equation.

EXAMPLE: Find the solution to $2x + x^2 = 8$.

SOLUTION:

Step 1: Write the equation in standard form

$$x^2 + 2x - 8 = 0$$

Step 2: Factor the left member

$$(x + 4)(x - 2) = 0$$

Step 3: Set each factor to zero

$$x + 4 = 0 \quad \text{and} \quad x - 2 = 0$$

Step 4: Solve each equation

$$
\begin{array}{ll}
x + 4 = 0 & x - 2 = 0 \\
x + 4 - 4 = 0 - 4 & x - 2 + 2 = 0 + 2 \\
x = -4 & x = 2
\end{array}
$$

Notice that there are two solutions. You can check each value.

$x = -4$:

$$2x + x^2 = 8$$

$$2(-4) + (-4)^2 = 8$$

$$-8 + 16 = 8$$

$$8 = 8$$

$x = 2$:

$$2x + x^2 = 8$$

$$2(2) + 2^2 = 8$$

$$4 + 4 = 8$$

$$8 = 8$$

EXAMPLE: Solve $2x^2 - 5 = 3x$.

SOLUTION:

Step 1: Write in standard form

$$2x^2 - 3x - 5 = 0$$

Step 2: Factor the left member

$$(2x - 5)(x + 1) = 0$$

Step 3: Set both factors equal to zero

$$2x - 5 = 0 \qquad x + 1 = 0$$

Step 4: Solve each equation

$$
\begin{aligned}
2x - 5 &= 0 & x + 1 &= 0 \\
2x - 5 + 5 &= 0 + 5 & x + 1 - 1 &= 0 - 1 \\
2x &= 5 & x &= -1 \\
\frac{2^1 x}{2^1} &= \frac{5}{2} & & \\
x &= \frac{5}{2} & &
\end{aligned}
$$

EXAMPLE: Solve $2x^2 = 18$.

SOLUTION:

Step 1: Write in standard form

$$2x^2 - 18 = 0$$

Step 2: Factor in the left member

$$2(x - 3)(x + 3) = 0$$

Step 3: Divide both sides by 2 and set both factors equal to zero

$$x - 3 = 0 \qquad x + 3 = 0$$

Step 4: Solve each equation

$$x - 3 + 3 = 0 + 3 \qquad x + 3 - 3 = 0 - 3$$
$$x = 3 \qquad\qquad x = -3$$

EXAMPLE: Solve $x^2 = 6x$.

SOLUTION:

Step 1: $x^2 - 6x = 0$

Step 2: $x(x - 6) = 0$

Step 3: $x = 0 \qquad x - 6 = 0$

Step 4: $\qquad\qquad x - 6 + 6 = 0 + 6$

$$x = +6$$

It should be noted that not all quadratic equations can be solved by factoring. However, for the purposes of this book, the solutions to the word problems in the next lesson can be solved using factoring.

Try These

Solve each:

1. $x^2 - 9x = -18$.

2. $x = x^2 - 20$.

3. $2x^2 + 9 = 9x$.

4. $6x^2 - 11x = 10$.

5. $x^2 = -8x$.

SOLUTIONS:

1. $$x^2 - 9x = -18$$

$$x^2 - 9x + 18 = 0$$

$$(x - 3)(x - 6) = 0$$

$x - 3 = 0$	$x - 6 = 0$
$x - 3 + 3 = 0 + 3$	$x - 6 + 6 = 0 + 6$
$x = 3$	$x = 6$

2. $$x = x^2 - 20$$

$$0 = x^2 - x - 20$$

$$0 = (x - 5)(x + 4)$$

$x - 5 = 0$	$x + 4 = 0$
$x - 5 + 5 = 0 + 5$	$x + 4 - 4 = 0 - 4$
$x = 5$	$x = -4$

3. $$2x^2 + 9 = 9x$$

$$2x^2 - 9x + 9 = 0$$

$$(2x - 3)(x - 3) = 0$$

$2x - 3 = 0$	$x - 3 = 0$
$2x - 3 + 3 = 0 + 3$	$x - 3 + 3 = 0 + 3$
$2x = 3$	$x = 3$

$$\frac{\cancel{2}^1 x}{\cancel{2}^1} = \frac{3}{2}$$

$$x = \frac{3}{2}$$

4. $$6x^2 - 11x = 10$$

$$6x^2 - 11x - 10 = 0$$

$$(2x - 5)(3x + 2) = 0$$

$$2x - 5 = 0 \qquad\qquad 3x + 2 = 0$$

$$2x - 5 + 5 = 0 + 5 \qquad 3x + 2 - 2 = 0 - 2$$

$$2x = 5 \qquad\qquad 3x = -2$$

$$\frac{2^1 x}{2^1} = \frac{5}{2} \qquad\qquad \frac{3^1 x}{3^1} = -\frac{2}{3}$$

$$x = \frac{5}{2} \qquad\qquad x = -\frac{2}{3}$$

5. $$x^2 = -8x$$

$$x^2 + 8x = 0$$

$$x(x + 8) = 0$$

$$x = 0 \qquad\qquad x + 8 = 0$$

$$x + 8 - 8 = 0 - 8$$

$$x = -8$$

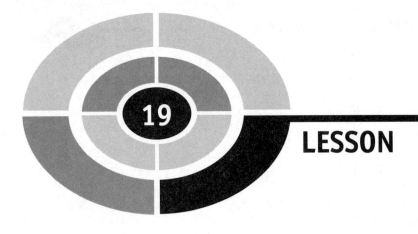

Solving Word Problems Using Quadratic Equations

Many problems in mathematics can be solved using a quadratic equation. The strategy you can use is

Step 1: Represent the unknown using x and the other unknown in terms of x.

Step 2: From the problem write expressions that are related to the unknown.

Step 3: Write the quadratic equation.

Step 4: Solve the quadratic equation for x.

Recall that a quadratic equation has two solutions. Both solutions can be answers to the problem; however, many times only one solution is

meaningful. This will be shown in some problems. In that case, disregard the solution that does not make sense. (*Note:* Sometimes the two solutions are equal to each other; in that case there is only one solution.)

EXAMPLE: If the sum of two numbers is 20 and the product of the two numbers is 36, find the numbers.

SOLUTION:

GOAL: You are being asked to find two numbers whose sum is 20 and whose product is 36.

STRATEGY: Let $x =$ one number
and $20 - x =$ the other number

If the product of the two numbers is 36, the equation is $x(20 - x) = 36$.

IMPLEMENTATION: Solve the equation:

$$x(20 - x) = 36$$

$$20x - x^2 = 36$$

$$0 = x^2 - 20x + 36$$

$$0 = (x - 18)(x - 2)$$

$x - 18 = 0$	$x - 2 = 0$
$x - 18 + 18 = 0 + 18$	$x - 2 + 2 = 0 + 2$
$x = 18$	$x = 2$

Hence the two numbers are 18 and 2.

EVALUATION: Check the facts of the problem. The sum $18 + 2$ is 20 and the product $18 \times 2 = 36$.

EXAMPLE: If the product of two consecutive numbers is 306, find the numbers.

SOLUTION:

GOAL: You are being asked to find two consecutive numbers whose product is 306.

STRATEGY: Let $x =$ the first number

and $x + 1 =$ the next number

The equation for the product is $x(x + 1) = 306$.

IMPLEMENTATION: Solve the equation:

$$x(x + 1) = 306$$

$$x^2 + x = 306$$

$$x^2 + x - 306 = 306 - 306$$

$$x^2 + x - 306 = 0$$

$$(x + 18)(x - 17) = 0$$

$x + 18 = 0$	$x - 17 = 0$
$x + 18 - 18 = 0 - 18$	$x - 17 + 17 = 0 + 17$
$x = -18$	$x = 17$
$x + 1 = -18 + 1 = -17$	$x + 1 = 17 + 1 = 18$

Hence the numbers are 17 and 18 or -17 and -18.

EVALUATION: Find the product: $17 \times 18 = 306$, and $-17 \times (-18) = 306$.

EXAMPLE: The sum of two numbers is 12. If the sum of their reciprocals is $\frac{3}{8}$, find the numbers.

SOLUTION:

GOAL: You are being asked to find two numbers whose sum is 12 and whose reciprocal sum is $\frac{3}{8}$.

STRATEGY: Let $x =$ one number

and $12 - x =$ the other number

The reciprocals are $\dfrac{1}{x}$ and $\dfrac{1}{12 - x}$.

Then the sum of the reciprocals is $\dfrac{1}{x} + \dfrac{1}{12 - x} = \dfrac{3}{8}$.

IMPLEMENTATION: Solve the equation:

$$\frac{1}{x} + \frac{1}{12 - x} = \frac{3}{8}$$

$$\text{L.C.D.} = 8x(12 - x)$$

$$8\cancel{x}(12 - x) \cdot \frac{1}{\cancel{x}} + 8x\cancel{(12 - x)}^1 \cdot \frac{1}{\cancel{(12 - x)}^1} = \cancel{8}^1 x(12 - x) \cdot \frac{3}{\cancel{8}^1}$$

$$8(12 - x) + 8x = 3x(12 - x)$$

$$96 - 8x + 8x = 36x - 3x^2$$

$$96 = 36x - 3x^2$$

$$3x^2 - 36x + 96 = 36x - 36x - 3x^2 + 3x^2$$

$$3x^2 - 36x + 96 = 0$$

Divide by 3: $\qquad x^2 - 12x + 32 = 0$

$$(x - 4)(x - 8) = 0$$

$$x - 4 = 0 \qquad\qquad\qquad x - 8 = 0$$

$$x - 4 + 4 = 0 + 4 \qquad\qquad x - 8 + 8 = 0 + 8$$

$$x = 4 \qquad\qquad\qquad\quad x = 8$$

EVALUATION: The sum of $4+8=12$. The sum of the reciprocals is $\frac{1}{4}+\frac{1}{8}=\frac{2}{8}+\frac{1}{8}=\frac{3}{8}$.

Try These

1. The square of a number plus 5 is equal to 86. Find the number.

2. One number is 6 more than another number, and the product of the two numbers is 91. Find the numbers.

3. If the product of two consecutive odd numbers is 255, find the numbers.

4. One number is 3 more than another number. If the square of the smaller number is 19 more than three times the larger number, find the numbers.

5. A larger pipe can fill a tank in 32 minutes less time than a smaller pipe. If they both are turned on at the same time, they can fill the tank in 30 minutes. How long does it take each pipe alone to fill the tank?

SOLUTIONS:

1. Let $x =$ the number

$$x^2 + 5 = 86$$

$$x^2 + 5 - 86 = 86 - 86$$

$$x^2 - 81 = 0$$

$$(x - 9)(x + 9) = 0$$

$$x + 9 = 0 \qquad\qquad x - 9 = 0$$

$$x + 9 - 9 = 0 - 9 \qquad x - 9 + 9 = 0$$

$$x = -9 \qquad\qquad x = 9$$

The answer is 9 and -9.

2. Let $x =$ one number

and $x + 6 =$ the other number

$$x(x + 6) = 91$$

$$x^2 + 6x = 91$$

$$x^2 + 6x - 91 = 91 - 91$$

$$x^2 + 6x - 91 = 0$$

$$(x + 13)(x - 7) = 0$$

$x + 13 = 0$	$x - 7 = 0$
$x + 13 - 13 = 0 - 13$	$x - 7 + 7 = 0 + 7$
$x = -13$	$x = 7$
$x + 6 = -13 + 6 = -7$	$x + 6 = 7 + 6 = 13$

The answers are -7 and -13, and 7 and 13.

3. Let $x =$ one number

and $x + 2 =$ the other number

$$x(x + 2) = 255$$

$$x^2 + 2x = 255$$

$$x^2 + 2x - 255 = 255 - 255$$

$$x^2 + 2x - 255 = 0$$

$$(x + 17)(x - 15) = 0$$

$x + 17 = 0$	$x - 15 = 0$
$x + 17 - 17 = 0 - 17$	$x - 15 + 15 = 0 + 15$
$x = -17$	$x = 15$
$x + 2 = -17 + 2 = -15$	$x + 2 = 15 + 2 = 17$

The answers are -15 and -17, and 15 and 17.

4. Let $x =$ the smaller number

and $x + 3 =$ the larger number

$$x^2 = 3(x + 3) + 19$$

$$x^2 = 3x + 9 + 19$$

$$x^2 = 3x + 28$$

$$x^2 - 3x - 28 = 3x - 3x + 28 - 28$$

$$x^2 - 3x - 28 = 0$$

$$(x - 7)(x + 4) = 0$$

$$x - 7 = 0 \qquad\qquad\qquad x + 4 = 0$$

$$x - 7 + 7 = 0 + 7 \qquad\qquad x + 4 - 4 = 0 - 4$$

$$x = 7 \qquad\qquad\qquad x = -4$$

$$x + 3 = 7 + 3 = 10 \qquad\qquad x + 3 = -4 + 3 = -1$$

The answers are 7 and 10, and -4 and -1.

5. Let $x =$ the time it takes the larger pipe to fill the tank

and $x + 32 =$ the time it takes the smaller pipe to fill the tank; then

$$\frac{30}{x} + \frac{30}{x + 32} = 1$$

$$\text{the L.C.D.} = x(x + 32)$$

$$\cancel{x}(x + 32) \cdot \frac{30}{\cancel{x}} + x\cancel{(x + 32)}^1 \cdot \frac{30}{\cancel{(x + 32)}^1} = x(x + 32)$$

$$30(x + 32) + 30x = x(x + 32)$$

$$30x + 960 + 30x = x^2 + 32x$$

$$60x + 960 = x^2 + 32x$$

$$60x - 60x + 960 = x^2 + 32x - 60x$$

$$960 = x^2 - 28x$$

$$960 - 960 = x^2 - 28x + 960$$

$$0 = x^2 - 28x + 960$$

$$0 = (x - 48)(x + 20)$$

$$(x - 48) = 0 \qquad\qquad x + 20 = 0$$

$$x - 48 + 48 = 0 + 48 \qquad x + 20 - 20 = 0 - 20$$

$$x = 48 \qquad\qquad\qquad x = -20$$

It will take 48 minutes. Ignore $x = -20$ since it is meaningless.

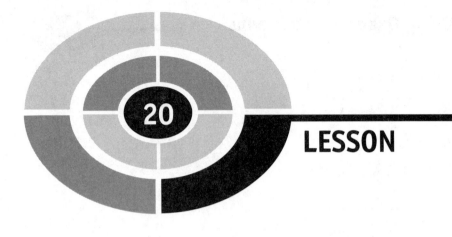

20

LESSON

Solving Word Problems in Geometry

Although word problems in geometry are for the most part different from those in algebra, many problems in geometry require algebra to solve them. It is not possible to show all the different types of problems that you will find in geometry, but a few of them will be explained here so that you can reach a basic understanding of how to use algebra to solve some of the problems found in geometry.

Each problem is based on a geometric principle or rule. The principles will be given here in each problem.

EXAMPLE: Find the measures of each angle of a triangle if the second angle is twice as large as the first angle and the third angle is equal to the measure of the second angle.

Geometric principle: The sum of the measures of the angles of a triangle is 180°.

SOLUTION:

GOAL: You are being asked to find the measures of the three angles of a triangle.

STRATEGY: Let $x =$ the measure of the first angle

and $2x =$ the measure of the second angle

and $2x =$ the measure of the third angle

See Figure 20-1.

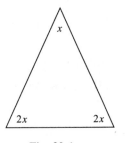

Fig. 20-1.

Since the sum of the measures of the angles of a triangle is 180°, the equation is

$x + 2x + 2x = 180°$.

IMPLEMENTATION: Solve the equation:

$$x + 2x + 2x = 180°$$
$$5x = 180°$$
$$\frac{\cancel{5}^1 x}{\cancel{5}^1} = \frac{180°}{5}$$

$$x = 36°$$
$$2x = 2 \cdot 36° = 72°$$

Hence the measures of each of the angles are 36°, 72°, and 72°.

EVALUATION: See if the sum of the angles is 180°.

$$36° + 72° + 72° = 180°$$

EXAMPLE: If the length of a rectangle is four times its width and the perimeter of the rectangle is 80 inches, find the measures of its length and width.

Geometric principle: The perimeter of a rectangle is $P = 2l + 2w$.

SOLUTION:

GOAL: You are being asked to find the length and width of a rectangle.

STRATEGY: Let x = the width of the rectangle
and $4x$ = the length of the rectangle

See Figure 20-2.

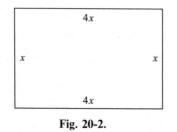

Fig. 20-2.

Since the formula for the perimeter of a rectangle is $P = 2l + 2w$, the equation is

$$2(4x) + (2x) = 80.$$

IMPLEMENTATION: Solve the equation:

$$2(4x) + (2x) = 80$$

$$8x + 2x = 80$$

$$10x = 80$$

$$\frac{\cancel{10}^{1}x}{\cancel{10}^{1}} = \frac{80}{10}$$

$$x = 8$$

$$4x = 4 \cdot 8 = 32$$

Hence the length is 32 inches and the width is 8 inches.

EVALUATION: Use the formula for perimeter and see if it is 80 inches.

$$P = 2l + 2w$$

$$P = 2(32) + 2(8)$$

$$= 64 + 16$$

$$= 80 \text{ inches}$$

EXAMPLE: The base of a triangle is 6 inches larger than its height. If the area of the triangle is 8 square inches, find the base and height of the triangle.

Geometric principle: The area of a triangle is $A = \frac{1}{2}bh$.

SOLUTION:

GOAL: You are being asked to find the measures of the base and the height.

STRATEGY: Let $x =$ the measure of the height

and $x + 6 =$ the measure of the base

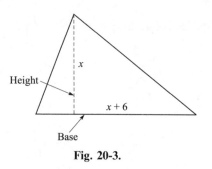

Fig. 20-3.

See Figure 20-3.

Since $A = \frac{1}{2}bh$, the equation is $8 = \frac{1}{2}(x+6)x$.

IMPLEMENTATION: Solve the equation:

$$8 = \frac{1}{2}(x+6)x$$

$$2 \cdot 8 = 2 \cdot \frac{1}{2}x(x+6)$$

$$16 = x(x+6)$$

$$16 = x^2 + 6x$$

$$16 - 16 = x^2 + 6x - 16$$

$$0 = x^2 + 6x - 16$$

$$0 = (x+8)(x-2)$$

$$x+8 = 0 \qquad\qquad x-2 = 0$$

$$x+8-8 = 0-8 \qquad\qquad x-2+2 = 0+2$$

$$x = -8 \qquad\qquad x = 2$$

In this case, we ignore $x = -8$ since a height cannot be a negative number.
The base is $x+6 = 2+6 = 8$ inches. Hence the height is 2 inches and the base is 8 inches.

EVALUATION: Find the area and see if it is 8 inches.

$$A = \frac{1}{2}bh$$

$$= \frac{1}{2}(8)(2)$$

$$= 8 \text{ square inches.}$$

Try These

1. The perimeter of a rectangle is 42 inches and the length is 3 times the width. Find its dimension. *Geometric principle*: The perimeter of a rectangle is $P = 2l + 2w$.

2. If two sides of a triangle are equal in length and the third side is 6 inches less than the length of one of the equal sides, find the length of the sides if the perimeter is 24 inches. *Geometric principle*: The perimeter of a triangle is equal to the sum of the lengths of its sides.

3. The perimeter of a rectangle is 52 inches. If the length is 10 inches less than twice the width, find its dimensions. $P = 2l + 2w$.

4. The sum of the angles of a triangle is 180°. If the second angle is twice as great as the first angle and the third angle is 20° less than the second angle, find the measures of the angles.

5. If an angle exceeds its complement by 25°, find its measure. *Geometric principle*: Complementary angles are adjacent angles whose sum is 90°.

6. If one angle of a triangle is 30° more than twice another angle, and the third angle is equal to the sum of the first two angles, find the measures of each angle. *Geometric principle*: The sum of the measures of the angles of a triangle is 180°.

7. The area of a rectangle is 60 square inches. The length is 4 inches longer than the width. Find its dimensions. *Geometric principle*: The area of a rectangle is $A = lw$.

8. The base of a triangle is 7 feet longer than its height. If its area is 30 square feet, find the measures of the base and height. Use $A = \frac{1}{2}bh$.

9. If the side of a large square is 6 times as long as the side of a smaller square and the area of the large square is 875 square inches larger than the area of the smaller square, find the length of the side of the smaller square. Use $A = s^2$.

10. If the area of a circle is 379.94 square inches, find the radius. Use $A = 3.14r^2$.

SOLUTIONS:

1.
$$\text{Let } x = \text{the width}$$
$$\text{and } x + 3 = \text{the length}$$
$$P = 2l + 2w$$
$$42 = 2(x + 3) + 2x$$
$$42 = 2x + 6 + 2x$$
$$42 = 4x + 6$$
$$42 - 6 = 4x + 6 - 6$$
$$36 = 4x$$
$$\frac{36}{4} = \frac{\cancel{4}^1 x}{\cancel{4}^1}$$
$$9 \text{ inches} = x$$
$$x + 3 = 9 + 3 = 12 \text{ inches}$$

The length is 12 inches and the width is 9 inches.

2.
$$\text{Let } x = \text{the length of one of the two equal sides}$$
$$\text{and } x - 6 = \text{the length of the third side}$$

$$x + x + x - 6 = 24$$
$$3x - 6 = 24$$
$$3x - 6 + 6 = 24 + 6$$
$$3x = 30$$
$$\frac{\cancel{3}^1 x}{\cancel{3}^1} = \frac{30}{3}$$
$$x = 10$$
$$x - 6 = 4$$

The lengths of the sides are 10 inches, 10 inches, and 4 inches.

3.   Let x = the width of the rectangle
and $2x - 10$ = the length of the rectangle
$$P = 2l + 2w$$
$$52 = 2(2x - 10) + 2x$$
$$52 = 4x - 20 + 2x$$
$$52 = 6x - 20$$
$$52 + 20 = 6x - 20 + 20$$
$$72 = 6x$$
$$\frac{72}{6} = \frac{\cancel{6}^1 x}{\cancel{6}^1}$$
$$12 \text{ inches} = x$$
$$2x - 10 = 2(12) - 10 = 14 \text{ inches}$$

The length is 14 inches and the width is 12 inches.

4.   Let x = the measure of one angle

$2x$ = the measure of the second angle

$2x - 20$ = the measure of the third angle

$$180 = x + 2x + 2x - 20$$

$$180 = 5x - 20$$

$$180 + 20 = 5x - 20 + 20$$

$$200 = 5x$$

$$\frac{200}{5} = \frac{\cancel{5}^1 x}{\cancel{5}^1}$$

$$40° = x$$

$$2x = 2(40) = 80°$$

$$2x - 20 = 2(40) - 20 = 60°$$

The measures of the three angles are 40°, 60°, and 80°.

5. Let x = the measure of one angle

$x + 25°$ = the measure of the other angle

$$x + x + 25° = 90 - 25$$

$$x + x + 25 - 25 = 90 - 25$$

$$2x = 65$$

$$\frac{\cancel{2}^1 x}{\cancel{2}^1} = \frac{65}{2}$$

$$x = 32.5°$$

$$x + 25° = 32.5 + 25 = 57.5°$$

6. Let x = the measure of one angle

$2x + 30$ = the measure of the second angle

$x + 2x + 30$ = the measure of the third angle

$$x + 2x + 30 + x + 2x + 30 = 180$$
$$6x + 60 = 180$$
$$6x + 60 - 60 = 180 - 60$$
$$6x = 120$$
$$\frac{\cancel{6}^{1}x}{\cancel{6}^{1}} = \frac{120}{6}$$
$$x = 20°$$
$$2x + 30 = 2(20) + 30 = 70°$$
$$20° + 70° = 90°$$

The measures of the angles are 20°, 70°, and 90°.

7. Let $x =$ the width of the rectangle

$x + 4 =$ the length of the rectangle

$$A = lw$$
$$60 = (x + 4)x$$
$$60 = x^2 + 4x$$
$$0 = x^2 + 4x - 60$$
$$0 = (x + 10)(x - 6)$$

$$x + 10 = 0 \qquad\qquad x - 6 = 0$$
$$x + 10 - 10 = 0 - 10 \qquad x - 6 + 6 = 0 + 6$$
$$x = -10 \qquad\qquad x = 6$$
$$x + 4 = 6 + 4 = 10$$

The length of the rectangle is 10 inches and the width is 6 inches.

8. Let $x =$ the measure of the height

and $x + 7 =$ the measure of the base

$$A = \frac{1}{2}bh$$

$$30 = \frac{1}{2}(x+7)x$$

$$30 \cdot 2 = \frac{\cancel{2}^{1}}{1} \cdot \frac{1}{\cancel{2}^{1}}(x+7)x$$

$$60 = x^2 + 7x$$

$$60 - 60 = x^2 + 7x - 60$$

$$0 = x^2 + 7x - 60$$

$$0 = (x+12)(x-5)$$

$x + 12 = 0$	$x - 5 = 0$
$x + 12 - 12 = 0 = 0 - 12$	$x - 5 + 5 = 0 + 5$
$x = -12$	$x = 5$

The height is 5 feet and the base is $5 + 7 = 12$ feet.

9. Let $x =$ the length of the side of the smaller square
and $6x =$ the length of the larger square; then

$$(6x)^2 - x^2 = 875$$

$$36x^2 - x^2 = 875$$

$$35x^2 = 875$$

$$\frac{\cancel{35}^{1} x^2}{\cancel{35}^{1}} = \frac{875}{35}$$

$$x^2 = 25$$

$$x = \sqrt{25} = 5 \text{ inches}$$

10.   Let x = the measure of the radius

$$A = 3.14r^2$$

$$379.94 = 3.14x^2$$

$$\frac{379.94}{3.14} = \frac{3.\cancel{14}^1 x^2}{3.\cancel{14}^1}$$

$$121 = x^2$$

$$\sqrt{121} = x$$

$$11 \text{ inches} = x$$

Quiz 5

1. Five hats and three scarves cost $99, while three hats and five scarves cost $85. Find the cost of one hat.
   - (a) $8
   - (b) $5
   - (c) $15
   - (d) $10

2. If a person can travel 20 miles upstream in 10 hours and the same distance downstream in 2.5 hours, find the rate of the current.
   - (a) 2 miles per hour
   - (b) 3 miles per hour
   - (c) 4 miles per hour
   - (d) 5 miles per hour

3. If the sum of two numbers is 43 and the difference is 7, find the larger number.
   - (a) 15
   - (b) 18
   - (c) 22
   - (d) 25

4. A person has $9000, part of which he invested at 3% interest and the rest of which he invested at 5%. If the yearly interest from the investment is $346, find the amount of money invested at 3%.
   (a) $5200
   (b) $6500
   (c) $3400
   (d) $4200

5. If the length and the width of a rectangle are increased by two inches, the area of the rectangle is 120 square inches. If the length and the width of a rectangle are decreased by two inches, the area of the rectangle is 48 square inches. Find the length of the rectangle.
   (a) 6 inches
   (b) 7 inches
   (c) 10 inches
   (d) 12 inches

6. The sum of Eli's age and Cecil's age is 60. Six years ago, Eli was three times as old as Cecil. Find Eli's age now.
   (a) 12
   (b) 18
   (c) 34
   (d) 42

7. A person plants two square plots for her garden. The total of the areas of both plots is 225 square feet. If the side of one plot is 3 feet longer than the side of the other one, find the length of the side of the smaller plot.
   (a) 8 feet
   (b) 9 feet
   (c) 12 feet
   (d) 14 feet

8. If the product of two consecutive even numbers is 168, find the larger one.
   (a) 8
   (b) 10
   (c) 12
   (d) 14

9. If the base of a triangle is 4 inches longer than its height and the area of the triangle is 48 square feet, find the length of the base. (Use $A = \frac{1}{2}bh$.)

   (a) 12 feet

   (b) 8 feet

   (c) 6 feet

   (d) 4 feet

10. Find the measure of the largest angle of a triangle if it is 3 times as large as the smallest angle and the second angle is 20° larger than the smallest angle. (The sum of the measures of the angles of a triangle is 180°.)

   (a) 52°

   (b) 80°

   (c) 96°

   (d) 108°

21

Solving Word Problems Using Other Strategies

In addition to being able to solve problems in prealgebra, algebra, and geometry, you can use some general problem-solving strategies to solve other real-world problems. These strategies can help you with problems found on standardized tests, in other subjects, and in everyday life.

These strategies are:

1. Make an organized list.
2. Guess and test.

3. Draw a picture.
4. Find a pattern.
5. Solve a simpler problem.
6. Work backwards.

Make an Organized List

When you use this strategy, you make an organized list of possible solutions and then systematically work out each one until the correct answer is found. Sometimes it helps to make the list in a table format.

EXAMPLE: A person has ten coins consisting of quarters and dimes. If the total amount of the change is $1.90, find the number of quarters and dimes he has.

SOLUTION:

GOAL: You are being asked to find the number of quarters and dimes the person has.

STRATEGY: As shown in Lesson 11, this problem can be solved using an equation; however, this problem can also be solved by making an organized list, as shown:

Quarters ($0.25)	Dimes ($0.10)	Amount
1	9	$1.15

One quarter and 9 dimes makes ten coins with a value of $1($0.25) + 9($0.10) = 1.15. Next try 2 quarters and 8 dimes and keep going until a sum of $1.90 is reached.

IMPLEMENTATION: Finish the list.

Quarters ($0.25)	Dimes ($0.10)	Amount
1	9	$1.15
2	8	$1.30
3	7	$1.45

4	6	$1.60
5	5	$1.75
6	4	$1.90

Hence, 6 quarters and 4 dimes are needed to get $1.90.

EVALUATION: Six quarters plus 4 dimes makes 10 coins whose value is $1.90. The answer is correct.

EXAMPLE: In a children's parade there are 12 children, some riding bicycles and some riding tricycles. If there is a total of 32 wheels, how many children are riding a bicycle and how many are riding a tricycle?

SOLUTION:

GOAL: You are being asked to find how many children are riding a bicycle and how many are riding a tricycle.

STRATEGY: You can make an organized list as shown:

Bicycles (2 wheels)	Tricycles (3 wheels)	Total wheels
1	11	35

The number of bicycles and tricycles must sum to 12 since there are 12 children.

IMPLEMENTATION: Continue the table until the correct answer (32 wheels) is found.

Bicycles (2 wheels)	Tricycles (3 wheels)	Total wheels
1	11	35
2	10	34
3	9	33
4	8	32

Hence there are 4 children riding bicycles and 8 children riding tricycles.

EVALUATION: Since $8 + 4 = 12$ and 4 bicycles have a total of 8 wheels and 8 tricycles have a total of 24 wheels, $8 + 24 = 32$.

Guess and Test

This strategy is similar to the previous one except you do not need to make a list. You simply take an educated guess at the solution and then try it out to see if it is correct.

EXAMPLE: The sum of the digits of a two-digit number is 11. If the digits are reversed, the new number is nine less than the original number.

SOLUTION:

GOAL: You are being asked to find a two-digit number.

STRATEGY: You can use the guess and test strategy. First guess some two-digit numbers such that the sum of the digits is 11. For example, 83, 74, 65, etc., meet this part of the solution. Then see if they meet the other condition of the problem.

IMPLEMENTATION:

Guess: 83; reverse the digits: 38; subtract: $83 - 38 = 45$

Guess: 74; reverse the digits: 47; subtract: $74 - 47 = 27$

Guess: 65; reverse the digits: 56; subtract: $65 - 56 = 9$

The number is 65.

EVALUATION: The sum of the digits is $6 + 5 = 11$, and the difference is $65 - 56 = 9$.

EXAMPLE: Each letter represents a digit from 0 through 9. Find the value of each letter so that the following is true:

$$
\begin{array}{r}
x \\
x \\
+\,x \\
\hline
wx
\end{array}
$$

SOLUTION:

GOAL: You are being asked to find what digits x and w represent.

STRATEGY: Use guess and test.

IMPLEMENTATION: Guess a few digits for x and see what works:

$$
\begin{array}{ccc}
x = 4\text{:} & x = 7\text{:} & x = 5\text{:} \\
\begin{array}{r} 4 \\ 4 \\ +4 \\ \hline 12 \end{array} &
\begin{array}{r} 7 \\ 7 \\ +7 \\ \hline 21 \end{array} &
\begin{array}{r} 5 \\ 5 \\ +5 \\ \hline 15 \end{array}
\end{array}
$$

Hence $x = 5$ and $w = 1$ is the correct answer.

EVALUATION: Notice that all the digits in the column are the same, that is, they are all the same number. You must add three single digit numbers and get the same number as the one's digit of the solution. There are only two possibilities: 0 and 5. Since the answer has two digits, 0 is disregarded.

Draw a Picture

Many times a problem can be solved using a picture, figure, or diagram. Also, drawing a picture can help you to determine which other strategy can be used to solve a problem.

EXAMPLE: Eight clothespins are placed on a clothesline at two-foot intervals. How far is it from the first one to the last one?

SOLUTION:

GOAL: You are being asked to find the distance from the first clothespin to the last one.

STRATEGY: Draw a figure and count the intervals between them; then multiply by two.

IMPLEMENTATION: Solve the problem. See Figure 21-1.

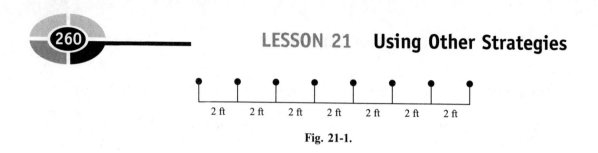

Fig. 21-1.

Since there are seven intervals, the distance between the first and last one is $7 \times 2 = 14$ feet.

EVALUATION: The figure shows that 14 feet is the correct answer.

EXAMPLE: Three coins are tossed; find the total number of ways they could fall.

SOLUTION:

GOAL: You are being asked to find the total number of ways 3 coins can fall.

STRATEGY: Draw a diagram showing the way each coin can land, that is, heads up or heads down.

IMPLEMENTATION: Each coin can land in two ways: heads or tails. See Figure 21-2.

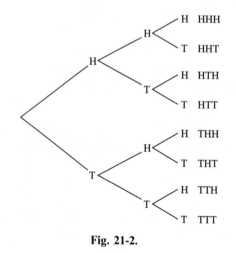

Fig. 21-2.

Hence there are 8 different possibilities:

HHH	THH
HHT	THT
HTH	TTH
HTT	TTT

EVALUATION: Since there are two ways each coin can fall, there are $2 \times 2 \times 2 = 8$ ways.

Find a Pattern

Many problems can be solved by recognizing that there is a pattern to the solution. Once the pattern is recognized, the solution can be obtained by generalizing from the pattern.

EXAMPLE: A wealthy person decided to pay an employee $1 for the first day's work, $3 for the second day's work, and $5 for the third day's work. How much did the employee earn for 30 days' work?

SOLUTION:

GOAL: You are being asked to find the amount the employee earned for a total of 30 days' work.

STRATEGY: You can make a table starting with the first day and continuing until you see a pattern.

IMPLEMENTATION:

Day	Amount earned	Total
1	1	1
2	3	4
3	5	9
4	7	16
5	9	25

Notice the relationship between the day and the total amount earned: $1^2 = 1$, $2^2 = 4$, $3^2 = 9$, $4^2 = 16$, $5^2 = 25$. Hence for 30 days' work, the employee would earn 30^2 or 900 dollars.

EVALUATION: You could check your answer by adding the first 30 odd numbers.

EXAMPLE: Find the sum of the digits for the answer to $(111,111,111)^2$.

SOLUTION:

GOAL: You are being asked to find the sum of the digits of the given number squared.

STRATEGY: Make a table starting with 1^2, 11^2, 111^2, etc. Find the sum of the digits and see if there is a pattern.

IMPLEMENTATION:

Number	Number squared	Sum of digits
1^2	1	1
11^2	121	4
111^2	12321	9
1111^2	1234321	16
11111^2	123454321	25

The pattern is similar to the one shown in the previous problem. If there are 9 digits in the number, the sum of the digits should be $9^2 = 81$.

EVALUATION: You could square the number and sum the digits.

Solve a Simpler Problem

To use this strategy, you should simplify the problem or make up a shorter, similar problem and figure out how to solve it. Then use the same strategy to solve the given problem.

EXAMPLE: If there are 8 people at a meeting and each person shakes every other person's hand once, how many handshakes occurred?

SOLUTION:

GOAL: You are being asked to find the total number of handshakes if everybody shakes everybody else's hand.

STRATEGY: Simplify the problem using, say, 4 people, and then try to solve it with 8 people.

IMPLEMENTATION: Assume the 4 people are A, B, C, and D. Then write the handshakes:

AB, AC, AD, BC, BD, CD

Hence, with 4 people, there would be 6 handshakes.

Now call the 8 people A, B, C, D, E, F, G, and H.

AB	AC	AD	AE	AF	AG	AH
BC	BD	BE	BF	BG	BH	
CD	CE	CF	CG	CH		
DE	DF	DG	DH			
EF	EG	EH				
FG	FH					
GH						

There would be 28 handshakes.

EVALUATION: You can solve the problem using a different strategy and see if you get the same answer.

Work Backwards

Some problems can be solved by starting at the end and working backwards to the beginning.

EXAMPLE: Tina spent $1.00 for parking and one-half of the remainder of her money in a department store. Then she spent $3.00 for lunch. Arriving back home, she found that she had $7.00 left. How much did she take to the store?

SOLUTION:

GOAL: You are being asked to find how much money Tina had before going shopping.

STRATEGY: Work backwards.

IMPLEMENTATION: Work forward first and then work backwards.

1. Spent: $1.00 on parking. Subtract $1.00.

2. Spent $\frac{1}{2}$ of the remainder in the department store. Divide by 2.

3. Spent $3.00 on lunch. Subtract $3.00.

4. Has $7.00 left.

Reversing the process:

4. $7.00

3. Add $3.00 $7.00 + $3.00 = $10.00

2. Multiply by 2 $10.00 × 2 = $20.00

1. Add $1.00 $20.00 + $1.00 = $21.00

Hence, she started out with $21.00.
Many times there is no single best strategy to solve a problem. You should remember that problems can be solved using different methods or a combination of methods.

Try These

Use one or more of the strategies shown in the lesson to solve each problem.

1. How many cuts are needed to cut a log into 5 pieces?

2. Each letter stands for a digit. All identical letters represent the same digit. Find the solution.

$$\begin{array}{r} PQ \\ + \underline{Q} \\ QP \end{array}$$

3. The sum of the digits of a two-digit number is 5. If 9 is subtracted from the number, the answer will be the original number with the digits reversed.

4. A person purchased 10 stamps of two denominations, $0.25 and $0.15. How many of each kind did the person purchase if the total is $2.10?

5. An 87-inch stick is broken into two pieces such that one piece is twice as long as the other. Find the length of each piece.

6. How many ways can a committee of three people be selected from four people?

7. Mortimer wants to shape up for football. He decides to cut back by eating two fewer candy bars each day for five days. During the five days, he ate a total of 30 candy bars. How many did he eat on the first day?

8. A father is three times as old as his son. In 12 years, he will be twice as old as his son. Find their present ages.

9. How many ways can 5 people line up in a row for a photograph?

10. Find the tallest person if Betty is shorter than Jan, Sue is taller than Betty, and Jan is shorter than Sue.

SOLUTIONS:

1. Strategy: Draw a picture. Four cuts are needed. See Figure 21-3.

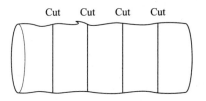

Cut   Cut   Cut   Cut

Fig. 21-3.

2.   Strategy: Guess and test. $89 + 9 = 98$.

3.   Strategy: Guess and test. $32 - 9 = 23$.

4.   Strategy: Make an organized list. 6 twenty-five cent stamps and 4 fifteen-cent stamps.

5.   Strategy: Guess and test. 29 inches and 58 inches.

6.   Strategy: Make an organized list. 4 ways.

7.   Strategy: Work backwards. 10 candy bars.

8.   Strategy: Make an organized list or guess and test. Father's age is 36, son's age is 12.

9.   Strategy: Solve a simpler problem. 120 ways.

10.   Strategy: Draw a picture. Sue.

Solving Word Problems in Probability

Probability deals with chance events, such as card games, slot machines, and lotteries, as well as insurance, investments, and weather forecasting. A **probability experiment** is a chance process which leads to well-defined outcomes. For example, when a die (singular for dice) is rolled, there are six possible well-defined outcomes. They are

1, 2, 3, 4, 5, 6

When a coin is flipped, there are two possible well-defined outcomes. They are

heads, tails

The set of all possible outcomes of a probability experiment is called the **sample space**. Each outcome in a sample space, unless otherwise noted, is considered equally likely, that is, it has the same chance of occurring. An **event** can consist of outcomes in the sample space. The basic definition of probability of an event is

$$P(E) = \frac{\text{Number of outcomes in event } E}{\text{Total number of outcomes in the sample space}}$$

The strategy when determining the probability of an event then is

1. Find the number of outcomes in event E.
2. Find the number of outcomes in the sample space.
3. Divide the first number by the second number to get a decimal or reduce the fraction if a fraction answer is desired.

EXAMPLE: A die is rolled; find the probability of getting an odd number.

SOLUTION:

GOAL: You are being asked to find the probability of getting an odd number.

STRATEGY: When a die is rolled, there are six outcomes in the sample space, and there are three outcomes in the event. That is, there are three odd numbers: 1, 3, and 5.

IMPLEMENTATION: $P(\text{odd number}) = \frac{3}{6} = \frac{1}{2}$ or 0.5.

EVALUATION: Since 1, 3, and 5 are half of the numbers in the sample space, the probability is correct.

When two coins are tossed, the sample space is

HH, HT, TH, TT

EXAMPLE: Two coins are tossed. Find the probability of getting a head and a tail in any order.

SOLUTION:

GOAL: You are being asked to find the probability of getting a head and a tail in any order.

STRATEGY: There are 4 outcomes in the sample space and two outcomes (HT, TH) in the event.

IMPLEMENTATION: $P = $ (one head and one tail) $= \dfrac{2}{4} = \dfrac{1}{2}$.

EVALUATION: Looking at the sample space, it is obvious that the probability of 2 choices from four outcomes is $\frac{1}{2}$.

When two dice are rolled, each die can have one of six outcomes. Therefore, there are $6 \times 6 = 36$ outcomes in the sample space. The outcomes can be arranged in ordered pairs such that the first number is the number of spots on the first die, and the second number in the pair is the number of spots on the second die. For example, the ordered pair $(2, 4)$ means a 2 came up on the first die and a 4 came up on the second die. Also, the sum of the numbers for this outcome is $2 + 4 = 6$.

The sample space for two dice is shown next:

$(1, 1)$ $(1, 2)$ $(1, 3)$ $(1, 4)$ $(1, 5)$ $(1, 6)$

$(2, 1)$ $(2, 2)$ $(2, 3)$ $(2, 4)$ $(2, 5)$ $(2, 6)$

$(3, 1)$ $(3, 2)$ $(3, 3)$ $(3, 4)$ $(3, 5)$ $(3, 6)$

$(4, 1)$ $(4, 2)$ $(4, 3)$ $(4, 4)$ $(4, 5)$ $(4, 6)$

$(5, 1)$ $(5, 2)$ $(5, 3)$ $(5, 4)$ $(5, 5)$ $(5, 6)$

$(6, 1)$ $(6, 2)$ $(6, 3)$ $(6, 4)$ $(6, 5)$ $(6, 6)$

EXAMPLE: Two dice are rolled; find the probability of getting a sum of 9.

SOLUTION:

GOAL: You are being asked to find the probability of getting a sum of 9.

STRATEGY: There are 36 outcomes in the sample space and four ways to get a sum of nine. They are $(3, 6)$, $(4, 5)$, $(5, 4)$, and $(6, 3)$.

IMPLEMENTATION: P(sum of 9) $= \dfrac{4}{36} = \dfrac{1}{9}$.

EVALUATION: Use the sample space to verify your answer.

EXAMPLE: Two dice are rolled; find the probability of getting a sum less than 4.

SOLUTION:

GOAL: You are being asked to find the probability of getting a sum less than 4.

STRATEGY: A sum less than 4 means a sum of 2 or 3. They are (1, 1), (1, 2), and (2, 1). Hence there are 3 ways to get a sum less than 4, and there are 36 outcomes in the sample space.

IMPLEMENTATION: P(sum less than 4) $= \dfrac{3}{36} = \dfrac{1}{12}$.

EVALUATION: Use the sample space to verify the answer.

Probability problems also use ordinary playing cards. In a deck of cards, there are 52 cards consisting of four suits: hearts and diamonds, which are red, and spades and clubs, which are black. In addition, there are 13 cards in each suit, ace through ten and a jack, a queen, and a king (called face cards).

EXAMPLE: A card is drawn from a deck. Find the probability that it is a seven.

SOLUTION:

GOAL: You are being asked to find the probability that the selected card is a seven.

STRATEGY: There are 52 outcomes in the sample space and 4 sevens.

IMPLEMENTATION: P(seven) $= \dfrac{4}{52} = \dfrac{1}{13}$.

EVALUATION: Use the sample space to verify the answer.

EXAMPLE: A card is selected from a deck; find the probability that it is a spade.

SOLUTION:

GOAL: You are being asked to find the probability of selecting a spade.

STRATEGY: There are 13 spades in a deck of 52 cards.

IMPLEMENTATION: $P(\text{spade}) = \dfrac{13}{52} = \dfrac{1}{4}$.

EVALUATION: Use the sample space to verify your answer.

The examples shown previously are examples of what is called **classical probability**. The next examples are from another area of probability called **empirical probability**. Empirical probability uses **frequency distributions**. Suppose that a bag of mixed candy contained 10 caramels, 8 peppermints, 4 chocolates, and 3 coconut creams. The sample space can be represented using a frequency distribution as shown.

Candy	Frequency
Caramels	10
Peppermints	8
Chocolates	4
Coconut creams	3
	25

This distribution can be used to solve probability problems.

EXAMPLE: Suppose a person selects a piece of candy from the bag; find the probability that it is a peppermint.

SOLUTION:

GOAL: You are being asked to find the probability that the piece of candy is a peppermint.

STRATEGY: There are 8 peppermints and a total of 25 pieces of candy, so the probability formula can be used.

IMPLEMENTATION: $P(\text{peppermint}) = \dfrac{8}{25}$.

EVALUATION: The answer can be verified by looking at the frequency distribution.

EXAMPLE: Using the same bag of candy, find the probability that a person selects a caramel or a chocolate.

SOLUTION:

GOAL: You are being asked to find the probability that the piece of candy is a caramel or a chocolate.

STRATEGY: There are 25 pieces of candy and there are 10 caramels and 4 chocolates.

IMPLEMENTATION: $P(\text{caramel or chocolate}) = \dfrac{10+4}{25} = \dfrac{14}{25}$.

EVALUATION: You can verify the answer by looking at the sample space.

EXAMPLE: In a classroom, there are 18 freshmen and 10 sophomores. If a student is selected at random to read a passage, find the probability that the student is a sophomore.

SOLUTION:

GOAL: You are being asked to find the probability that the student is a sophomore.

STRATEGY: There are a total of 28 students in the class and 10 are sophomores.

IMPLEMENTATION: $P(\text{sophomore}) = \dfrac{10}{28} = \dfrac{5}{14}$.

EVALUATION: The answer can be verified by looking at the distribution.

Try These

1. A single die is rolled once; find the probability of getting
   (a) a five
   (b) a number greater than four
   (c) a number less than six
   (d) a number greater than six

2. Two dice are rolled; find the probability of getting
    (a)   a sum of 8
    (b)   doubles
    (c)   a sum greater than 10
    (d)   a sum less than 5

3. A card is drawn from a deck; find the probability of getting
    (a)   the six of diamonds
    (b)   a queen
    (c)   a heart
    (d)   a club or a spade
    (e)   a red card

4. A couple has three children; find the probability that the children are
    (a)   all boys
    (b)   all boys or all girls
    (c)   exactly two boys and one girl

5. Two dice are rolled; find the probability of getting a sum of seven or eleven.

6. In a cooler there are 8 cans of cola and 4 cans of ginger ale. If a person selects a can of soda without looking at it, find the probability that it is a can of ginger ale.

7. In a box of Popsicles there are 3 orange ones, 5 cherry ones, 4 grape ones, and 4 banana ones. If a child selects one without looking at it, find the probability that it is grape.

8. Fifty tickets are sold for a prize. If you buy 5 tickets, find the probability that you will win the prize.

9. In an elementary school there are 87 girls and 49 boys. If one student is selected by a drawing to attend a conference, find the probability that the student is a girl.

10. If an organization wishes to schedule a national conference and they select a state by drawing it out of a hat, find the probability that the state's name begins with the letter M.

SOLUTIONS:

1.
    (a)   There are six outcomes in the sample space and one outcome is a 5; therefore, $P(5) = \frac{1}{6}$.

(b) There are six outcomes in the sample space and there are two outcomes that are greater than 4, that is, 5 and 6; hence, P(a number greater than 4) $=\frac{2}{6}=\frac{1}{3}$.

(c) There are six outcomes in the sample space and five numbers less than six; hence, P(a number less than six) $=\frac{5}{6}$.

(d) There are six outcomes in the sample space and no numbers are greater than six; hence, P(a number greater than six) $=\frac{0}{6}=0$.

2.

(a) There are 36 outcomes in the sample space and there are five ways to get a sum of 8: (2, 6), (3, 5), (4, 4), (5, 3), and (6, 2); hence, P(sum of 8) $=\frac{5}{36}$.

(b) There are 36 outcomes in the sample space and six ways to get doubles: (1, 1), (2, 2), (3, 3), (4, 4), (5, 5), and (6, 6); hence P(doubles) $=\frac{6}{36}=\frac{1}{6}$.

(c) There are 36 outcomes in the sample space and there are 3 ways of getting a sum greater than ten, that is, a sum of 11 or 12: (5, 6), (6, 5), and (6, 6). Hence, P(sum greater than 10) $=\frac{3}{36}=\frac{1}{12}$.

(d) There are 36 outcomes in the sample space and six ways to get a sum of 4, 3, or 2: (1, 3), (2, 2), (3, 1), (1, 2), (2, 1), and (1, 1). Hence, P(sum less than 3) $=\frac{6}{36}=\frac{1}{6}$.

3.

(a) There are 52 outcomes in the sample space and one six of diamonds; hence, P(six of diamonds) $=\frac{1}{52}$.

(b) There are 52 outcomes in the sample space and four queens; hence, P(queen) $=\frac{4}{52}=\frac{1}{13}$.

(c) There are 52 outcomes in the sample space and 13 hearts; hence, P(heart) $=\frac{13}{52}=\frac{1}{4}$.

(d) There are 52 outcomes in the sample space and 13 clubs and 13 spades; hence, P(club or spade) $=\frac{13+13}{52}=\frac{26}{52}=\frac{1}{2}$.

(e) There are 52 outcomes in the sample space and 26 red cards (13 diamonds and 13 hearts); hence, P(red card) $=\frac{26}{52}=\frac{1}{2}$.

4. The sample space for three children is

BBB GGB
BBG GBG
BGB BGG
GBB GGG

(a) There are 8 outcomes in the sample space and one way to get all boys: BBB; hence, $P(\text{all boys}) = \frac{1}{8}$.

(b) There are 8 outcomes in the sample space and two ways to get all boys or all girls: BBB and GGG; hence, $P(\text{all boys or all girls}) = \frac{2}{8} = \frac{1}{4}$.

(c) There are 8 outcomes in the sample space and three ways to get two boys and a girl: BBG, BGB, GBB: hence $P(\text{exactly 2 boys and 1 girl}) = \frac{3}{8}$.

5. There are 36 outcomes in the sample space and six ways to get a 7 and two ways to get an 11; hence, $P(7 \text{ or } 11) = \frac{6+2}{36} = \frac{8}{36} = \frac{2}{9}$.

6. There are $8 + 4 = 12$ cans in the cooler and 4 of them are ginger ale; hence $P(\text{ginger ale}) = \frac{4}{12} = \frac{1}{3}$.

7. There are $3 + 5 + 4 + 4 = 16$ Popsicles in the box and 4 are grape; hence, $P(\text{grape}) = \frac{4}{16} = \frac{1}{4}$.

8. There are 50 tickets and you have 5, so the probability of winning is $\frac{5}{50} = \frac{1}{10}$.

9. There are a total of $87 + 49 = 136$ students and 87 are girls; hence, $P(\text{girl}) = \frac{87}{136}$.

10. There are 50 states and 8 of them begin with the letter M: Maine, Maryland, Massachusetts, Michigan, Minnesota, Mississippi, Missouri, and Montana. Hence, $P(\text{state that begins with M}) = \frac{8}{50} = \frac{4}{25}$.

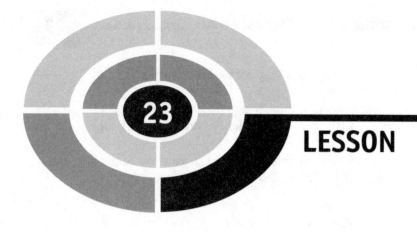

Solving Word Problems in Statistics

Statistics is the science of conducting studies to collect, organize, analyze, summarize, and draw conclusions from data. The **data** can be numbers such as weights, temperatures, test scores, etc., or observations such as colors of automobiles, political affiliations, etc. A group of data values collected for a particular study is called a **data set**. Statistics is used in almost all fields of human endeavor.

In statistics, there are three commonly used measures of average. They are the mean, median, and mode.

The **mean** is the sum of the data values divided by the total number of data values.

EXAMPLE: Find the mean of 8, 15, 19, 24, and 14.

SOLUTION:

GOAL: You are being asked to find the mean for the given data set.

STRATEGY: Add the values and divide the sum by 5 (there are 5 data values).

IMPLEMENTATION:

$$8 + 15 + 19 + 24 + 14 = 80$$
$$80 \div 5 = 16$$

The mean is 16.

EVALUATION: The mean will fall between the lowest and highest values; most of the time, somewhere near the middle of the values.

The **median** is a value which falls in the center of the data set. You must first arrange the data in order from the smallest data value to the largest data value.

EXAMPLE: Find the median for 8, 15, 19, 24, and 14.

SOLUTION:

GOAL: You are being asked to find the median for the given data set.

STRATEGY: Arrange the data values in order and find the middle value.

IMPLEMENTATION:

8, 14, 15, 19, 24

Since 15 is the middle value, the median is 15.

EVALUATION: Check to see if the data values are arranged correctly; then make sure you have found the middle value.

If the number of data values is odd, as in the previous example, the median will be one of the values; however, if the number of data values is even, the median will fall halfway between the middle two values, as shown in the next example.

EXAMPLE: Find the median for 56, 82, 34, 63, 68, and 72.

SOLUTION:

GOAL: You are being asked to find the median for the given data set.

STRATEGY: Arrange the data in order; then find the middle point.

IMPLEMENTATION:

34, 56, 63, 68, 72, 82

The middle of the data is halfway between 63 and 68; hence, the median is

$$\frac{63 + 68}{2} = \frac{131}{2} = 65.5$$

EVALUATION: Recheck the problem.

The third measure of average is called the mode. The **mode** is the data value which occurs most often.

EXAMPLE: Find the mode of 24, 17, 32, 17, 15, and 19.

SOLUTION:

GOAL: You are being asked to find the mode for the given data set.

STRATEGY: Find the value which occurs most often.

IMPLEMENTATION:

It is helpful although not necessary to arrange the data in order:

15, 17, 17, 19, 24, 32

Since 17 occurs twice, and that is more often than any other number, 17 is the mode.

EVALUATION: The answer is obvious.

EXAMPLE: Find the mode for 2, 3, 5, 5, 5, 6, 6, 8, 8, 8, 12, and 15.

SOLUTION:

GOAL: You are being asked to find the mode for the given data set.

STRATEGY: Analyze the data and see what value occurs most often.

IMPLEMENTATION: In this case, the values of 5 and 8 occur three times. Hence, the data has two modes. They are 5 and 8.

EVALUATION: The answer is obvious.

EXAMPLE: Find the mode for 56, 82, 34, 63, 68, and 72.

SOLUTION:

GOAL: You are being asked to find the mode for the given data set.

STRATEGY: Find the data value which occurs most often.

IMPLEMENTATION: In this case, each data value occurs only once. Hence, we say that there is no mode.

Two things should be noted:

1. The mode of a data set can be a single value, more than one value, or no value at all.
2. The mean, median, and mode for a data set, in most cases, will not be equal.

In addition to the measures of average, statisticians also use measures of variation to describe a data set. The two most often used measures of variation are the range and the standard deviation. These measures measure the spread of the data about the mean. Loosely speaking, the larger the range or standard deviation, the more variable or spread out the data is in the set.

The **range** is found by subtracting the smallest data value from the largest data value.

EXAMPLE: Find the range for 8, 15, 19, 24, and 14.

SOLUTION:

GOAL: You are being asked to find the range for the given data set.

STRATEGY: Subtract the smallest data value from the largest data value in the set.

IMPLEMENTATION: The smallest data value is 8, and the largest data value is 24, so the range is $24 - 8 = 16$.

EVALUATION: Redo the problem.

The range is a rough estimate of variation, so statisticians also use what is called the standard deviation. The standard deviation can be computed by using the following procedure:

1. Find the mean for the data set.
2. Subtract the mean from each value in the data set.
3. Square the differences.
4. Find the sum of the squares.
5. Divide the sum by $n - 1$ where n is the number of data values.
6. Take the square root of the answer. (You may need a calculator for this step.)

EXAMPLE: Find the standard deviation: 10, 12, 15, 18, and 20.

SOLUTION:

GOAL: You are being asked to find the standard deviation for the given data set.

STRATEGY: Use the procedure given.

IMPLEMENTATION:

1. Find the mean:

$$10 + 12 + 15 + 18 + 20 = 75$$
$$75 \div 5 = 15$$

2. Subtract the mean from each data value:

$$10 - 15 = -5$$
$$12 - 15 = -3$$
$$15 - 15 = 0$$
$$18 - 15 = 3$$
$$20 - 15 = 5$$

3. Square the answers:

$$(-5)^2 = 25$$
$$(-3)^2 = 9$$
$$0^2 = 0$$
$$3^2 = 9$$
$$5^2 = 25$$

4. Find the sum of the squares:

$$25 + 9 + 0 + 9 + 25 = 68$$

5. Divide the sum by $n - 1$, where $n = 5$ and $n - 1 = 5 - 1 = 4$:

$$68 \div 4 = 17$$

6. Find the square root of 17:

$$\sqrt{17} = 4.12 \text{ (two decimal places)}$$

The standard deviation is 4.12.

EVALUATION: The standard deviation can be estimated by dividing the range by 4. In this case, the range is 10. Thus, $10 \div 4 = 2.5$. Since this is only a rough estimate, we are in the ball park.

Roughly speaking, most of the data values will usually fall between two standard deviations of the mean.

Try These

For the data set 32, 25, 18, 18, and 17, find each:

1. The mean.
2. The median.
3. The mode.
4. The range.
5. The standard deviation.

SOLUTIONS:

1. $32 + 25 + 18 + 18 + 17 = 110$ $110 \div 5 = 22$ Mean $= 22$

2. 17, 18, 18, 25, 32   The middle value is 18; hence, the median is 18.

3. The value which occurs most often is 18, so the mode is 18.

4. The range is $32 - 17 = 15$.

5. Follow these steps:

    Find the mean. It is 22, as found in answer 1.

    Subtract the mean from each data value:

    $$32 - 22 = 10$$
    $$25 - 22 = 3$$
    $$18 - 22 = -4$$
    $$18 - 22 = -4$$
    $$17 - 22 = -5$$

    Square the differences:

    $$10^2 = 100$$
    $$3^2 = 9$$
    $$(-4)^2 = 16$$
    $$(-4)^2 = 16$$
    $$(-5)^2 = 25$$

    Find the sum of the differences:

    $$100 + 9 + 16 + 16 + 25 = 166$$

    Divide the sum by 4:

    $$\frac{166}{4} = 41.5$$

    Find the square root of 41.5:

    $$\sqrt{41.5} = 6.44$$

Quiz 6

1. A single die is rolled; the probability of getting a one is
   (a) $\frac{1}{3}$
   (b) $\frac{1}{6}$
   (c) 0
   (d) 1

2. A card is selected from a deck; the probability of getting a black card is
   (a) $\frac{1}{4}$
   (b) $\frac{1}{52}$
   (c) $\frac{1}{13}$
   (d) $\frac{1}{2}$

3. Two dice are rolled; the probability of getting a sum of 11 is
   (a) $\frac{1}{18}$
   (b) $\frac{1}{36}$
   (c) $\frac{11}{36}$
   (d) $\frac{1}{2}$

4. Find the mean of 8, 24, 15, 19, 32, and 10.
   (a) 16
   (b) 17
   (c) 20
   (d) 18

5. Find the median of 6, 10, 13, 7, and 20.
   (a) 10
   (b) 11.2
   (c) 13
   (d) 15

6. Find the median of 23, 26, 28, and 35.
   (a) 26
   (b) 27
   (c) 28
   (d) 29

7. Find the mode of 25, 16, 42, 48, and 33.
   (a) 32.8
   (b) 33
   (c) 42
   (d) no mode

8. Find the range of 5, 12, 14, 16, 20, and 19.
   (a) 15
   (b) 14
   (c) 16
   (d) 4

9. Find the standard deviation of 9, 12, 18, 13, 17.
   (a) 2.8
   (b) 3.7
   (c) 13.7
   (d) 2

10. To estimate the standard deviation, the range should be divided by what number?
    (a) 1
    (b) 2
    (c) 3
    (d) 4

Final Exam

1. The height of St. Helen's volcano is 8363 feet, and the height of the Augustine volcano is 3999 feet. How much higher is the height of St. Helen's volcano?
   (a) 4364 feet
   (b) 12,362 feet
   (c) 5346 feet
   (d) 2316 feet

2. Find the total of the areas of the Mediterranean Sea (969,100 square miles) and the Red Sea (174,900 square miles).
   (a) 794,200 square miles
   (b) 1,144,000 square miles
   (c) 873,2000 square miles
   (d) 1,382,600 square miles

3. If a person saves $75 a month, how much will the person have saved in a year?
   (a) $87

    (b) $450
    (c) $750
    (d) $900

4. How many boxes are needed to package 500 calculators if 20 calculators can fit in a box?
    (a) 520
    (b) 480
    (c) 25
    (d) 10,000

5. A person traveled from her home to a local mall, a distance of $9\frac{3}{4}$ miles. Then she went to her health club, a distance of $5\frac{2}{3}$ miles from the mall. How far did she travel in all?
    (a) $55\frac{1}{4}$ miles
    (b) $1\frac{49}{68}$ miles
    (c) $15\frac{5}{12}$ miles
    (d) $4\frac{1}{12}$ miles

6. A limousine service charges $50 plus 85 cents per mile to rent a limousine. How much does a person pay for a 27-mile trip?
    (a) $22.95
    (b) $27.05
    (c) $39.45
    (d) $72.95

7. A clerk sold $1\frac{3}{4}$ pounds of licorice, $2\frac{5}{8}$ pounds of peppermints, and $3\frac{1}{2}$ pounds of caramels. How many pounds of candy were sold in all?
    (a) $7\frac{3}{8}$ pounds
    (b) $7\frac{7}{8}$ pounds
    (c) $6\frac{3}{4}$ pounds
    (d) $6\frac{5}{8}$ pounds

8. A person traveled 344.4 miles on 14 gallons of gasoline. How many miles per gallon did the person get?
    (a) 23.8 miles per hour
    (b) 25.2 miles per hour
    (c) 28.4 miles per hour
    (d) 24.6 miles per hour

9. An automobile depreciates 25% of its value after the first year. How much is a car that originally costs $22,275 worth after one year?
   (a) $16,706.25
   (b) $5568.75
   (c) $17,521.75
   (d) $4823.25

10. A person receives a 6% raise. Find this new salary if he earns $26,500.
    (a) $1496
    (b) $28,090
    (c) $1590
    (d) $27,300

11. What is the selling price of a computer if the sales tax is $27.48 and the rate is 6%?
    (a) $1548.80
    (b) $372
    (c) $876.40
    (d) $458

12. If 5 gallons of driveway sealer will cover 300 square feet, how many gallons of sealer will be needed to cover a driveway that is 480 square feet?
    (a) 6 gallons
    (b) 8 gallons
    (c) 7.5 gallons
    (d) 8.5 gallons

13. Sally bought 20 cards and paid $40.25. If some of the cards cost $1.75 and the rest cost $2.50, how many of the $2.50 cards did she buy?
    (a) 7
    (b) 9
    (c) 11
    (d) 13

14. Sam is eight years older than his brother. Seven years ago, Sam was twice as old as his brother. Find Sam's age.
    (a) 7
    (b) 15
    (c) 16
    (d) 23

15. One pipe can fill a pool in 12 hours and another pipe can drain the pool in 21 hours. If both pipes are opened, how long will it take to fill the pool?
   (a) 24 hours
   (b) 26 hours
   (c) 28 hours
   (d) 30 hours

16. A lever is 8 feet long. Where should the fulcrum be placed in order to balance 50 pounds at one end and 150 pounds from the other end?
   (a) 6 feet from the 150 pounds
   (b) 2 feet from the 150 pounds
   (c) 2.5 feet from the 150 pounds
   (d) 5.5 feet from the 150 pounds

17. If the product of two consecutive odd numbers is 195, find the larger number.
   (a) 5
   (b) 13
   (c) 15
   (d) 25

18. If the length of a rectangular field is 9 yards more than its width and the area of the field is 400 square yards, find the length of the field.
   (a) 20 yards
   (b) 25 yards
   (c) 15 yards
   (d) 16 yards

19. A person has two accounts at a bank. One account pays 6.5% interest and the other pays 4%. If the total investment is $18,400 and the total interest is $968.50, find the amount of money the person has invested at 4%.
   (a) $10,400
   (b) $9100
   (c) $8600
   (d) $9300

20. A child's bank contains 89 coins consisting of dimes and quarters only. Find the number of dimes it contains if the total amount in the bank is $16.70.
    (a) 41
    (b) 37
    (c) 34
    (d) 52

21. The sum of the digits of a two-digit number is 15. If the digits are reversed, the new number is 9 more than the original number. Find the number.
    (a) 87
    (b) 69
    (c) 96
    (d) 78

22. A person in a motorboat traveled 48 miles upstream in 6 hours. The same trip downstream took 3 hours. Find the rate of the current.
    (a) 4 miles per hour
    (b) 8 miles per hour
    (c) 10 miles per hour
    (d) 12 miles per hour

23. In a two-digit number, the ten's digit is 3 more than the unit's digit. If the digits are reversed, the new number is 27 less than the original number. Find the number.
    (a) 63
    (b) 74
    (c) 97
    (d) 52

24. An airplane took 20 hours to fly a distance of 1500 miles, flying against the wind. If the return trip took 12 hours flying with the wind, find the speed of the wind.
    (a) 15 miles per hour
    (b) 20 miles per hour
    (c) 25 miles per hour
    (d) 35 miles per hour

25. How far will an automobile travel in $2\frac{3}{4}$ hours at a speed of 42 miles per hour? Use $D = RT$.
    (a) $110\frac{3}{4}$ miles
    (b) $112\frac{1}{2}$ miles
    (c) $115\frac{1}{2}$ miles
    (d) $121\frac{1}{4}$ miles

26. Find the interest on a loan of $8000 at 6% for 5 years. Use $I = PRT$.
    (a) $4800
    (b) $2400
    (c) $3200
    (d) $3000

27. Find the perimeter of a rectangle whose length is 27 feet and whose width is 9 feet. Use $P = 2l + 2w$.
    (a) 36 feet
    (b) 243 square feet
    (c) 80 feet
    (d) 72 feet

28. Find the distance an object falls in 9 seconds. Use $d = \frac{1}{2}(32)t^2$.
    (a) 1296 feet
    (b) 144 feet
    (c) 96 feet
    (d) 960 feet

29. One angle of a triangle is 40°. If another angle is 5° more than the third angle, find the measure of the third angle. The sum of the measures of the angles of a triangle is 180°.
    (a) 67.5°
    (b) 72.5°
    (c) 87.5°
    (d) 82.5°

30. A factory has 352 employees. If there are 16 more males than females, find the number of females employed by the factory.
    (a) 192
    (b) 184
    (c) 168
    (d) 176

31. When three coins are tossed, the probability of getting 3 tails is

    (a) $\dfrac{1}{4}$

    (b) $\dfrac{1}{2}$

    (c) $\dfrac{1}{8}$

    (d) $\dfrac{7}{8}$

32. When a single die is rolled, the probability of getting a seven is

    (a) $\dfrac{1}{6}$

    (b) 0

    (c) 1

    (d) $\dfrac{7}{6}$

33. When a card is selected from a deck, the probability of getting a 3 or a club is

    (a) $\dfrac{1}{13}$

    (b) $\dfrac{1}{4}$

    (c) $\dfrac{17}{52}$

    (d) $\dfrac{4}{13}$

34. In a meeting, there are six faculty members and two principals. If a person is selected at random to be the group leader, the probability that the leader is a principal is

    (a) $\dfrac{1}{8}$

    (b) $\dfrac{3}{4}$

    (c) $\dfrac{3}{8}$

    (d) $\dfrac{1}{4}$

35. When a die is rolled, the probability of getting a number less than seven is

    (a) $\dfrac{1}{6}$

    (b) $\dfrac{5}{6}$

    (c) $\dfrac{2}{3}$

    (d) 1

36. Find the mean of 9, 15, 18, 14, 23, 32, 13, and 20.
    (a) 14
    (b) 15
    (c) 18
    (d) 16.5

37. Find the median of 9, 15, 18, 14, 23, 32, 45, and 5.
    (a) 10.5
    (b) 16.5
    (c) 18.5
    (d) 14.5

38. Find the median of 21, 32, 26, 15, and 12.
    (a) 21.2
    (b) 26
    (c) 15
    (d) 21

39. Find the range of 106, 112, 98, 143, 152, and 127.
    (a) 54
    (b) 21
    (c) 38
    (d) 15

40. Find the standard deviation of 7, 19, 32, 15, and 9.
    (a) 98.8
    (b) 10.32
    (c) 9.94
    (d) 6.25

Answers to Quizzes and Final Exam

ANSWERS TO QUIZZES

	Quiz 1	Quiz 2	Quiz 3	Quiz 4	Quiz 5	Quiz 6
1.	(c)	(d)	(b)	(c)	(c)	(b)
2.	(b)	(b)	(d)	(a)	(b)	(d)
3.	(d)	(a)	(a)	(d)	(d)	(a)
4.	(a)	(c)	(d)	(b)	(a)	(d)
5.	(c)	(d)	(b)	(c)	(c)	(a)
6.	(a)	(c)	(b)	(a)	(d)	(b)
7.	(c)	(a)	(d)	(d)	(b)	(d)
8.	(b)	(b)	(c)	(b)	(d)	(a)
9.	(a)	(d)	(d)	(b)	(a)	(b)
10.	(d)	(c)	(b)	(d)	(c)	(d)

ANSWERS TO FINAL EXAM

1. (a)	11. (d)	21. (d)	31. (c)
2. (b)	12. (b)	22. (a)	32. (b)
3. (d)	13. (a)	23. (d)	33. (d)
4. (c)	14. (d)	24. (c)	34. (d)
5. (c)	15. (c)	25. (c)	35. (d)
6. (d)	16. (b)	26. (b)	36. (c)
7. (b)	17. (c)	27. (d)	37. (b)
8. (d)	18. (b)	28. (a)	38. (d)
9. (a)	19. (b)	29. (a)	39. (a)
10. (b)	20. (b)	30. (b)	40. (c)

Supplement: Suggestions for Success in Mathematics

1. Be sure to attend every class. If you know ahead of time that you will be absent, tell your instructor and get the assignment. If it is an emergency absence, get the assignment from another student. Try to do the problems before the next class. If possible, get the class notes from another student.

2. Read the material in the textbook several times. Write down or underline all definitions, rules, and symbols. Try to do the sample problems.

3. Do all assigned homework as soon as possible before the next class. Concentrate on mathematics only. Get all of your materials before you start doing your homework. Make sure you write the assignment on the top of your homework. Read the directions. Copy each problem on your homework paper. Make sure that you have copied it correctly. Do *not* use scratch paper. Work out each problem in detail and do not skip steps. Write neatly and large enough. Check the answer with the one in the back of the book or rework the problem again. If you did not get the correct answer, try to find your mistake or start over. Don't look for shortcuts because they do not always work. Write down any questions you have and ask your instructor or another student at the next class period. If you are having difficulty with the problem, consult your textbook and notes. Don't give up too quickly.

4. Always review before each exam. You can usually find a review or chapter test at the end of each chapter in the book. If not, you can make up your own review by selecting several problems from each section in the book to try. If you can't get the correct answer, ask the teacher or another student to help you before the exam. If you have made study cards, review them.

5. On the day of the test, arrive early. Look over your notes and study cards. Bring all necessary materials such as pencils, protractor, calculator, textbook, etc., to class. When you get the test, look over the entire test before you get started. Read the directions. Work the problems that you know how to do first. Do not spend too much time on any one problem. After you have finished the test, if time allows, check each problem. When you get the test back, check your mistakes and study the types of problems that you have missed because similar problems may be on the final exam.

6. If you have difficulty with mathematics, arrange for a tutor. Some schools have learning centers where you can receive free tutoring.

7. Finally, make sure that you are in the correct class. You cannot skip math classes. Mathematics is sequential in nature. What you learn today, you will use tomorrow. What you learn in one course you will use in the next course.

Good luck!

INDEX

ABOUT THE AUTHOR

Allan G. Bluman taught mathematics and statistics in high school, college, and graduate school for 39 years. He received his Ed.D. from the University of Pittsburgh and has written three mathematics textbooks published by McGraw-Hill. Dr Bluman is the recipient of an "Apple for the Teacher" award for bringing excellence to the learning environment and the "Most Successful Revision of a Textbook" award from McGraw-Hill. His biographical record appears in *Who's Who in American Education*, 5th edition.